Robust Optimization

Robust Optimization:
World's Best Practices for Developing Winning Vehicles

Subir Chowdhury
Shin Taguchi

ASI Consulting Group, LLC, USA

This edition first published 2016
© 2016 Subir Chowdhury, Shin Taguchi and ASI Consulting Group, LLC. All rights reserved.

Registered office
John Wiley & Sons Ltd, The Atrium, Southern Gate, Chichester, West Sussex, PO19 8SQ, United Kingdom

For details of our global editorial offices, for customer services and for information about how to apply for permission to reuse the copyright material in this book please see our website at www.wiley.com.

Library of Congress Cataloging-in-Publication Data

Names: Chowdhury, Subir, author. | Taguchi, Shin, author.
Title: Robust optimization : world's best practices for developing winning
    vehicles / Subir Chowdhury, Shin Taguchi.
Description: Chichester, West Sussex, United Kingdom : John Wiley & Sons,
    Inc., [2016] | Includes bibliographical references and index.
Identifiers: LCCN 2015035260 | ISBN 9781119212126 (cloth)
Subjects: LCSH: Motor vehicles–Design and construction. | Robust
    optimization. | Manufacturing processes.
Classification: LCC TL240 .C435 2016 | DDC 629.2/31–dc23 LC record available at
http://lccn.loc.gov/2015035260

A catalogue record for this book is available from the British Library.

ISBN: 9781119212126

Set in 10/12.5 pt PalatinoLTStd-Roman by Thomson Digital, Noida, India

1   2016

In memory of our teachers, colleagues, friends:
Genichi Taguchi
Yuin Wu

Contents

Preface

What is Robust Optimization? Put simply, it's a method to improve robustness using low-cost variations of a single, conceptual design. Jim Pratt, a former Vice-President of ITT, once said that using Robust Optimization on manufacturing processes was like "picking gold up off the floor!" Robust Optimization uses Robust Assessment to estimate the robustness of low-cost combinations of design parameter (control factor) values with a single conceptual design in order to discover the most robust combination of design parameter (control factor) values. A design parameter is called a control factor in Robust Optimization. A control factor value is a quantitative or qualitative level of a variable contained within the selected conceptual design. In mechanical designs, dimensions, radii, and material properties are typical control factors. In electrical designs, resistance, impedance, and capacitance are frequent control factors. In chemical processes, temperatures, rate of temperature change, pressures, pressure rate, reagents, and catalysts are common control factors.

In product or process development, Robust Optimization occurs after system (conceptual) design is complete and before the conceptual design is adjusted to meet requirements. Robust Optimization is best conducted before the requirements are defined.

Robust Optimization as a first step discovers a robust, low-cost combination of control factor values. This combination may not meet or even be close to meeting the design requirements. However, Robust Optimization as a second step discovers how to adjust that low-cost combination of control factor values so that the product can meet requirements. That very stable combination of control factor values is adjusted or "tuned" so that the product or process can easily meet requirements or specifications.

Selection of design concept that is robust is critical. Robust Optimization and assessment allow us to evaluate how robust the new concept is quickly so that we can try many concepts. We don't want to pass poor concepts through design gates. We need to detect bad design early. If we are going to fail, it is better to fail early so we can move on to the next concept quickly.

The benefits of Robust Optimization include such things as faster product development cycles; faster launch cycles; fewer manufacturing problems; fewer field problems; lower-cost, higher performing products and processes; and lower warranty. All these benefits can be realized if Engineering and product development leadership of automotive and manufacturing organizations leverage the power of using Robust Optimization as a competitive weapon.

The overarching benefit, however, is to create a group of technical employees with skills that produce results which dazzle your customers and the general public. Become the organization that produces the highest quality products at the lowest cost. Become the organization that is the first or second choice of every top-ranked engineering graduate in the world. Become the organization that is featured in the trade magazines as the innovator with rock solid quality and a winning value proposition for your customers.

The main objective of this book is to introduce engineering executives and leaders to the technical management strategy of Robust Optimization. In the first three chapters, we will discuss what the strategy entails, Eight Steps for Robust Optimization and Robust Assessment, and how to lead it in a technical organization with an implementation strategy. Another objective of this book is to demonstrate the application of Robust Optimization to automotive applications using real-life case studies from leading automotive organizations.

In this book, we define Robust Optimization and demonstrate how these techniques can be applied to build into manufacturing organizations especially with automotive industry applications; that Robust Optimization creates the flexibility that minimizes product development cost, reduces product time-to-market, and increases overall productivity. For the past 40 years thousands of companies throughout the world have been using the methodology 'Robust Optimization' of the late Dr. Genichi Taguchi, the quality pioneer, and have obtained positive results. Some organizations have integrated this new powerful methodology into their corporate culture. The benefits these organizations have achieved are phenomenal.

In this groundbreaking book, we have organized 19 successfully implemented best case studies from automotive original equipment manufacturers such as General Motors, Ford, Chrysler, Nissan, Isuzu, and Mazda as well as automotive suppliers like Bosch, Delphi, and Alps Electric. We have been working for past decades with all types of clients in automotive, manufacturing, healthcare, food, aerospace and other industries. Our firm ASI Consulting Group, LLC, headquartered in Bingham Farms, Michigan, USA, is very fortunate in that clients have been continuously putting their trust in our team. Most importantly, clients share their success stories at our annual client conferences. We are also fortunate to have other organizations from Europe and Asia to attend our annual conference and present their success stories on the application of Robust Optimization. ASI has thousands of case studies in its database and it is therefore impossible to feature all case studies in this book. However, we have included some of the very best of these case studies, keeping in mind the variety of applications of Robust Optimization.

There are many books available, mostly in English and Japanese, which include some automotive case studies. However, there is no book to date which presents *best practices* from around the globe on Robust Optimization in the automotive industry, and manufacturing industry in general. This is the first book to focus on the automotive application of Robust Optimization. In the organizations where Robust Optimization is extremely successful, senior leadership has an understanding of the significant impact of the applications and therefore support for implementation is highly encouraged or mandatory. In the United States, Asia, and Europe, hundreds of organizations have unfortunately been using Robust Optimization incorrectly, but those that have used the methodology correctly have been saving millions of dollars. Those organizations that have not been utilizing this powerful method may have not being doing so because of their lack of management understanding and misconceptions about its complexity. This book will therefore be a must read for any engineering manager or engineer because of its ability to clarify these generalized misconceptions.

This is the first book that features case studies from all four critical areas of Robust Optimization of an automotive organization:

1. Vehicle Level Optimization
2. Subsystems Level Optimization by Original Equipment Manufacturers (OEMs)
3. Subsystems Level Optimization by Suppliers
4. Manufacturing Process Optimization.

We also hope that this book provides direct learning techniques to the vast variety of industries and educational institutions and that it provides a formula for *instant knowledge* on areas that apply to the reader and his/her organization.

This book is for engineers and managers who are working in the design, product, manufacturing, mechanical, electrical, process, and quality areas; all levels of management especially in the product development area; and research and development personnel and consultants. Almost all the case studies featured in this book make it suitable as a training and education guide, as well as serving both students and teachers in engineering colleges. We strongly feel that all libraries with technical sections will greatly benefit from having this book in their collection.

Subir Chowdhury
Shin Taguchi
July 31, 2015

Acknowledgments

The authors gratefully acknowledge the efforts of all who assisted in the completion of this book:

- To all of the contributors and their organizations for sharing their successful case studies.
- To the ASI Consulting Group, LLC and all its employees and partners worldwide.
- To our colleagues and friends, Alan Wu, Brad Walker, Michael O'Ship, Matt Gajda, Michael Holbrook, Brian Bartnick, Bill Eureka, Jay Eleswarpu, Joe Smith, and Francois Pelka for effectively promoting Robust Optimization each day.
- To our colleague Jodi Caldwell for her hard work on the preparation of the manuscript.
- To our two retired colleagues and friends, Jim Quinlan and Barry Bebb for enriching us over the past two decades.
- To Paul Petralia, our editor at Wiley, for his dedication and guidance to make the book better.
- To Liz Wingett, our project editor at Wiley, for her hard work making the book published on time.
- To Martin Noble, our copy editor at Wiley, for his dedicated efforts toward refining the manuscript.
- To Anne Hunt, Associate Commissioning Editor at Wiley for her work on the book.
- To Sandra Grayson, Associate Book Editor, for her work on the book.
- Finally, this book never would have been materialized without the continuous support of our wonderful wives Malini Chowdhury and Junko Taguchi.

About the Authors

Subir Chowdhury has been a thought leader in quality management strategy and methodology for more than 20 years. Currently Chairman and CEO of ASI Consulting Group, LLC, he leads Six Sigma and Quality Leadership implementation, and consulting and training efforts. Subir's work has earned him numerous awards and recognition. *The New York Times* cited
him as a "leading quality expert"; *BusinessWeek* hailed him as the "Quality Prophet." *The Conference Board Review* described him as "an excitable, enthusiastic evangelist for quality."

Subir has worked with many organizations across diverse industries including manufacturing, healthcare, food, and nonprofit organizations. His client list includes major global corporations and industrial leaders such as American Axle, Berger Health Systems, Bosch, Caterpillar, Daewoo, Delphi Automotive Systems, Fiat-Chrysler Automotive, Ford, General Motors, Hyundai Motor Company, ITT Industries, Johns Manville, Kaplan Professional, Kia Motors, Leader Dogs for the Blind, Loral Space Systems, Make It Right Foundation, Mark IV Automotive, Procter & Gamble, State of Michigan, Thomson Multimedia, TRW, Volkswagen, Xerox, and more. Under Subir's leadership, ASI Consulting Group has helped hundreds of clients around the world save billions of dollars in recovered productivity and increased revenues.

Subir is the author of 14 books, including the international bestseller *The Power of Six Sigma* (Dearborn Trade, 2001), which has sold more than a million copies worldwide and been translated into more than 20 languages. *Design for Six Sigma* (Kaplan Professional, 2002) was the first book to popularize the

"DFSS" concept. With quality pioneer Dr. Genichi Taguchi, Subir co-authored two technical bestsellers *Robust Engineering* (McGraw Hill, 1999) and *Taguchi's Quality Engineering Handbook* (Wiley, 2005).

His book, the critically acclaimed *The Ice Cream Maker* (Random House Doubleday, 2005) introduced LEO® (Listen, Enrich, Optimize), a flexible management strategy that brings the concept of quality to every member of an organization. The book was formally recognized and distributed to every member of the 109th Congress. The LEO process continues to be implemented in many organizations. His most recent book, *The Power of LEO* (McGraw-Hill, 2011) was an *Inc.* Magazine bestseller. A follow-up to *The Ice Cream Maker*, the book shows organizations how the LEO methodology can be integrated into a complete quality management system.

London, UK based *Thinkers50* named Subir as one of the "50 Most Influential Management Thinkers in the World" in 2011, 2013 and 2015. Subir is a recipient of the Society of Manufacturing Engineers' Gold Medal, the Society of Automotive Engineers' (SAE) Henry Ford II Distinguished Award for excellence in Automotive Engineering and the American Society of Quality's first Philip Crosby Medal for authoring the most influential book on Quality. The US Department of Homeland Security presented the "Outstanding American by Choice Award" to Subir for his contributions to the field of quality and management.

In 2014, the University of California at Berkeley established the *Subir & Malini Chowdhury Center for Bangladesh Studies*. The Center will award graduate fellowships, scholarships, and research grants that focus on ways to improve the quality of life for the people of Bangladesh.

Each year the *Subir Chowdhury Fellowship on Quality and Economics* is awarded by both Harvard University and London School of Economics and Political Science to a doctoral student to research and study the impact of quality in the economic advancement of a nation. The SAE International established the "Subir Chowdhury Medal of Quality Leadership," an annual award that recognizes those individuals who promote innovation and expand the impact of quality in mobility engineering, design and manufacturing.

Subir received his undergraduate degree in Aeronautical Engineering from the Indian Institute of Technology (IIT), Kharagpur, India and his graduate degree in Industrial Management from Central Michigan University, Mt. Pleasant, Michigan. He has received Distinguished Alumnus Awards from both universities, as well as an honorary doctorate of engineering from the Michigan Technological University.

Subir lives with his wife, Malini and two children, Anandi and Anish, in Los Angeles, California.

Shin Taguchi is Chief Technical Officer (CTO) for ASI Consulting Group, LLC. He is a Master Black Belt in Six Sigma and Design for Six Sigma (DFSS) and was one of the world authorities in developing the DFSS program at ASI-CG, an internationally recognized training and consulting organization, dedicated to improving the competitive position of industries. He is the son of Dr. Genichi Taguchi, developer of new engineering approaches for robust technology that have saved American industry billions of dollars.

Over the last thirty years, Shin has trained more than 60 000 engineers around the world in quality engineering, product/process optimization, and robust design techniques, Mahalanobis-Taguchi System, known as Taguchi Methods™. Some of the many clients he has helped to make products and processes Robust include: Ford Motor Company, General Motors, Delphi Automotive Systems, Fiat-Chrysler Automotive, ITT, Kodak, Lexmark, Goodyear Tire & Rubber, General Electric, Miller Brewing, The Budd Company, Westinghouse, NASA, Texas Instruments, Xerox, Hyundai Motor Company, TRW and many others. In 1996, Shin developed and started to teach a Taguchi Certification Course. Over 360 people have graduated to date from this ongoing 16-day master certification course.

Shin is a Fellow of the Royal Statistical Society in London, and is a member of the Institute of Industrial Engineering (IIE) and the American Society for Quality (ASQ); Shin is a member of the Quality Control Research Group of the Japanese Standards Association (JSA) and Quality Engineering Society of Japan. He is an editor of the Quality Engineering Forum Technical Journal and was awarded the Craig Award for the best technical paper presented at the annual conference of the ASQ. Shin has been featured in the media through a number of national and international forums, including *Fortune* Magazine and *Actionline* (a publication of AIAG). Shin co-authored *Robust Engineering*, published by McGraw Hill in 1999. He has given presentations and workshops at numerous conferences, including ASQ, ASME, SME, SAE, and IIE. He is also a Master Black Belt for Design for Six Sigma (DFSS).

Shin holds a Bachelor of Science degree in Industrial Engineering and Statistics from the University of Michigan and trains and consults with many major corporations worldwide.

Shin lives with his wife, Junko and three children, Hana, Yumi and Miki, in West Bloomfield, Michigan.

1

Introduction to Robust Optimization

The automotive industry is very dynamic and the product is continuously changing. The competition is so cut-throat that it is becoming increasingly important to deliver quality products at all times. The customers are demanding the highest quality product at a cheaper price. Robust optimization is the mantra for automotive product development organizations both for original equipment manufacturers (OEMs) and their suppliers, especially in this competitive environment. Dr. Genichi Taguchi's *Robust Optimization* idea is simply revolutionary. To practice robust optimization correctly, product development and manufacturing organizations need to change the way they work, the way work is done needs to change, the way work is managed needs to change, knowledge and skills need to change, the way organizations are led needs to change. Obviously, all of these take time. Not accepting this reality will be more devastating in the future for any organization that wants to win customers' hearts by consistently delivering highest quality products.

Dr. Genichi Taguchi talked about quality as loss to society and how that loss is estimated using a "Quality Loss Function." He talked about robustness – the functional stability of products or processes in the face of ubiquitous variation in the usage conditions (noise factors). He talked about a product development process involving system, parameter and tolerance design steps.

Robust Optimization: World's Best Practices for Developing Winning Vehicles,
First Edition. Subir Chowdhury and Shin Taguchi.
© 2016 Subir Chowdhury, Shin Taguchi, and ASI Consulting Group, LLC.
Published 2016 by John Wiley & Sons, Ltd.

He suggested that engineers focus less on meeting requirements and more on discovering combinations of design variable values that (1) stabilize the function and (2) control the adjustment or "tuning" of that function. He talked about ideal functions.

Dr. Taguchi asked engineers and engineering leadership to look at technical work in an entirely different light.

What happened?

Well, since the word "quality" was part of the "Quality Loss Function," the quality experts in the organization took over that concept.

Robustness sounded like product performance in the field. So robustness was delegated to the reliability and validation engineers. Noise factors seemed similar to best case and worst case conditions, so that, too, was a good fit to reliability and validation engineering.

His recommended product development system sounded a lot like existing concurrent engineering and optimization methodologies. System engineers looked at Dr. Taguchi's comments and said, "We already do this – there's nothing new here!"

Parameter design was seen as setting design variable values at levels that met requirements in all conditions. Since parameter design borrowed orthogonal arrays from design of experiments, Taguchi's methods were often seen as a form of Design of Experiment. In most engineering organizations, Designed Experiments were organized by a quality expert when engineering had a problem. Parameter design was delegated to quality and product engineering. Often, an experiment was conducted only if a problem of sufficient magnitude presented itself. Taguchi's parameter design methods were roundly criticized by statisticians for, among many other things, a lack of statistical rigor. Even today, "Taguchi Designs" remain a subset of most statistical computer programs. A subset only "recommended" for preliminary, screening experiments.

1.1 What Is Quality as Loss?

One of our client engineers once had a car with a noisy transmission. He took it to the dealer because the noise bothered him. The dealer attached a machine to the transmission. It printed out a report.

"Your transmission is within specification," the dealer said.

There was nothing more to be done. He drove the car for a couple of years. He was glad when he could replace it with a new one. He never bought that brand of car again – even though their transmission was in specification. The dealer's machine and the printout said so.

Dr. Genichi Taguchi defines quality as "Quality may be assessed as the minimum loss imparted by the product to society from the time the product is shipped." The larger the loss, the poorer is the quality. This kind of thinking says that there is a difference among products even if they are within specification.

The "ideal" amount of noise from an automotive transmission is zero (yes, it's impossible to achieve). As the noise from the transmission increases it will bother some people more than others. But when the noise bothers someone enough, he or she will suffer a loss. They have to take the time to drive to the dealer and wait while the service technician conducts a diagnosis. There will be a dollar value for his time. The drive, diagnosis and report out will take about two hours. Two hours at that time in this person's life is probably worth about $250. Is that the total loss? What about the company's loss of a future sale? How much is that worth? What is the profit the company would make from the sale? The loss suffered by the company who made the noisy transmission is certainly more than $250.

If an automotive manufacturer makes a very, very noisy transmission, a customer might insist that it be replaced. It doesn't matter if the transmission is in or out of specification. The customer wants it replaced. The total loss to society is probably around $3500 (including customer inconvenience). It doesn't matter whether the transmission is under warranty or not. If under warranty, the manufacturer pays; if not, the customer pays. Either way "society" is out $3500 for each transmission that is so noisy it needs to be replaced.

Using this type of data, the quality in regards to audible noise of any transmission can be estimated. The actual amount of audible noise in decibels could be placed along the bottom axis. Dr. Genichi Taguchi is suggesting that every transmission that makes any noise at all contributes a slight amount of loss to society.

The redefinition of quality that you, as the technical leader of your organization, need to embrace is that producing parts within specification is absolutely necessary. However, only producing parts that meet requirements is no longer competitive.

For long-term success in the marketplace, we must focus on producing low-cost products that lower the loss to society. The average dollars lost by society due to audible transmission noise can be estimated for the transmissions made by your company versus the transmissions made by your competition. The long-term competitive position of your company correlates well with such estimates. Products with lower quality loss to society do better over time in the market. Where do your products rate?

While automobiles provide value to society such as transportation and pleasure of driving, automobiles are producing significant amounts of losses. Those losses include emissions, global warming, and automobile accidents. Dr. Taguchi always dreamt about accident-free automobiles and automobiles that clean air.

1.2   What Is Robustness?

What is robustness? You may have to dust off some of your old textbooks (or go online), but you can do it. The ideas aren't that complicated for a technically trained person like you. Let's define robustness as *the ability of a product or process to function consistently as the surrounding uncontrollable or uncontrolled factors vary.*

An example is the power window system in the driver's side door of your car. Does it perform today as well as it did the day you took delivery of it? On an extremely cold morning? On a hot summer day? When you are sitting in the car with the motor off? At 50 mph? Has the window ever stopped working entirely?

If two window systems are being compared, the more robust window system is the one that performs most consistently over a large number of cycles, at low and high temperatures, when running on battery power, or when the car is moving a high speed.

Higher robustness means that a product will last longer in the field, that is, in the hands of the customer. No matter how old the vehicle, no customer should have to awkwardly open the door of her car on a cold winter day to pay and pick up her order at the drive-through window. Only window systems with high levels of robustness can meet that requirement.

Robustness is easy to understand. We appreciate the chain of coffee stores that provides a cup of coffee with consistent taste, aroma, and temperature, regardless of whether we buy it in Seattle or Shanghai. We gravitate toward products that perform consistently over a long useful life. A carpenter needs a circular saw that will last for years of hard use after being thrown into the back of a pickup truck. The expensive two-fuel stove in our kitchen shouldn't have the control panel fail in the first month we own it.

One common misunderstanding about robustness is that more expensive products tend to be more robust. We think that we have to pay for robustness. But is a luxury brand car more robust than a small traditional sedan of one-quarter of the price? In many regards, probably not. More importantly, robust optimization provides methods by which high robustness can be achieved at low cost.

1.3 What Is Robust Assessment?

Robustness is a measurement, not a requirement to be reached. Robustness is only meaningful in comparison. Is my product more or less robust than my competitor's? By how much? Is the new design more or less robust than the old design? By how much? The measure or robustness is the signal-to-noise ratio (S/N ratio). The higher the S/N ratio, the more robust the product or process.

Use the creativity of your people to develop methods to assess (estimate) the robustness of your products in 15 minutes! Usually no more than six measurements are needed to estimate robustness. Most companies that use these ideas strategically develop special fixtures to help engineers estimate robustness quickly and efficiently.

After learning and applying Robust Assessment, an Engineering Vice President at Ricoh said, "From now on, our assessment on a paper handling system will take only two sheets of paper." At Nissan, a robust assessment technique was developed that takes only 15 minutes to assess robustness of a power window system with a high confidence level.

John Elter, a former VP of Engineering at Xerox, said that engineering labs used to be filled with prototype copy machines running continuously for life test and to estimate failure rate. After Robust Assessment, they are filled with jigs and fixtures to measure functions and robustness; functions include paper feeding, toner dispensing, toner charging, toner transfer, fusing, etc.

1.4 What Is Robust Optimization?

Robust optimization, a concept as familiar as it is misunderstood, will be clarified in this chapter. We conduct robust optimization by following the two-step process: (1) Minimize variability in the product or process, and (2) adjust the output to hit the target. In other words, first optimize performance to get the best out of the concept selected, then adjust the output to the target value to confirm whether all the requirements are met. The better the concept can perform, the greater our chances to meet all requirements. In the first step we try to kill many birds with one stone, that is, to meet many requirements by doing only one thing. How is that possible?

We start by identifying the ideal function, which will be determined by the basic physics of the system, be it a product or process. In either case, the design will be evaluated by the basic physics of the system. When evaluating a product or a manufacturing process, the ideal function is defined based on energy transformation from the input to the output. For example, for a car to go faster, the driver presses down on the gas pedal, and that energy is transformed to increased speed by sending gas through a fuel line to the engine, where it is burned, and finally to the wheels, which turn faster.

When designing a process, energy is not transformed, as in the design of a product, but information is. Take the invoicing process, for example. The supplier sends the company an invoice, and that information starts a chain of events that transforms the information into various forms of record-keeping and results, finally, in a check being sent to the supplier.

In either case, we first define what the ideal function for that particular product or process would look like; then we seek a design that will minimize the variability of the transformation of energy or information, depending on what we are trying to optimize.

We concentrate on the transformation of energy or information because all problems, including defects, failures, and poor reliability, are symptoms of variability in the transformation of energy or information. By optimizing that transformation – taking out virtually all sources of "friction" or noise along the way – we strive to meet all the requirements at once.

To understand fully this revolutionary approach, let's first review how quality control has traditionally worked. Virtually since the advent of commerce, a "good" or acceptable product or process has been defined simply as one that meets the standards set by the company. But here's the critical weakness to the old way of thinking: It has always been assumed that *any* product or process that falls *anywhere* in the acceptable range is equal to any other that falls within that range.

Picture the old conveyer belt, where the products roll along the line one by one until they get to the end, where an inspector wearing goggles and a white coat looks at each one and tosses them either into the "acceptable" bin or the "reject" bin. In that case, there are no other distinctions made among the finished products, just "okay" or "bad."

If you were to ask that old-school inspector what separates the worst "okay" specimen from the best reject – in other words, the ones very close to the cutoff line – he'd probably say something like, "It's a hair difference, but you've got to draw the line somewhere." But the inspector treats all acceptable samples the same: He just tosses them in the "okay" bin, and the same with the rejects. Even though he knows there are a million shades of gray in the output, he separates them all into black or white.

Now if you asked a typical consumer of that product if there was any difference between a sample that barely met the standards to make into the "okay" bin and one that was perfect, she'd say, "Yes, absolutely. You can easily tell the difference between these two."

The difference between the inspector's and the customer's viewpoints can be clarified further with the following analogy: If both people were playing darts, the inspector would only notice whether or not the dart hit the dartboard, not caring if it landed near the edge of the board or right on the bull's-eye. But to the customer, there would be a world of difference between the dart that landed on the board's edge and the one that pierced the bull's-eye. Although she certainly wouldn't want any dart not good enough to hit the board, she would still greatly prefer the bull's-eye to the one just an inch inside the board's edge. The point is: With the old way of inspecting products, the manufacturer or service provider made no distinctions among acceptable outputs, but the consumer almost always did, which made the company out of step with the customer's observations and desire.

This dissonance between these two perspectives demonstrates that the traditional view of quality – "good enough!" – is not good enough for remaining competitive in the modern economy. Instead of just barely meeting the lowest possible specifications, we need to hit the bull's-eye. The way to do that is to replace the oversimplified over/under bar with a more sophisticated bull's-eye design, where the goal is not merely to make acceptable products, but to reduce the spread of darts around the target.

The same is also true on the other side of the mark. In the old system, once you meet the specification that was that. No point going past it. But even if we're already doing a good job on a particular specification, we need to look into whether we can do it better and, if so, what it would cost. Would improving pay off?

Robust optimization requires you to free your employees – and your imaginations – to achieve the optimum performance by focusing on the energy/information transformation described earlier. This notion of having no ceiling is important, not just as a business concept, but psychologically as well. The IRS, of course, tells you how much to pay in taxes, and virtually no one ever pays extra. Most taxpayers do their best to pay as little as legally possible. Charities, on the other hand, never tell their donors what to pay – which might explain why Americans are by far the most generous citizens around the world in terms of charitable giving.

The point is simple: Don't give any employee, team, or project an upper limit. Let them optimize and *maximize* the design for robustness. See what's possible, and take advantage of the best performances you can produce! Let the sky be the limit and watch what your people can do! A limitless environment is a very inspiring place to work.

The next big question is: Once the energy/information transformation is optimized, is the design's performance greater than required? If so, you've got some decisions to make. Let's examine two extreme cases.

When the optimum performance exceeds the requirements, you have plenty of opportunities to reduce real cost. For example, you can use the added value in other ways, by using cheaper materials, increased tolerances, or by speeding up the process. The objective of robust optimization is to improve performance without increasing costs. Once you can achieve that, you can take advantage of the opportunities that cost reduction can create.

On the flip side, if the optimum performance comes in below the requirements, it's time to rethink the concept and come up with something better. The problem is that, in most corporate cultures, it is very difficult to abandon a concept because so many people have already spent so much time and effort on the project.

But this is where leadership comes in. Despite the heartbreak of letting an idea go, if it's not good enough, it's not good enough. So instead of spending good money on a doomed project and fighting fires later, it's best to cut your losses, reject the concept (salvaging the best ideas, if any), and move on to the next one, instead of locking yourself into a method of production that's never going to give you the results you want. Thus, it is extremely important to detect poor designs and reject them at the early stages of development.

Dr. Genichi Taguchi has built a model based on this concept that demonstrates the impact that variations from the target have on profits and costs. As the function of the product or process deviates from the target – either above or below it – the quality of the function is compromised. This in turn results in higher losses. The further from the target, the greater the monetary losses will be.

1.4.1 Noise Factors

The bugaboos that create the wiggles in the products and processes we create can be separated into the following general categories:

- manufacturing, material, and assembly vitiations;
- environmental influences (not ecological, but atmospheric);
- customer causes;
- deterioration, aging, and wear;
- neighboring subsystems.

This list will become especially important to us when we look at parameter design for robust optimization, whose stated purpose is to minimize the system's sensitivity to these sources of variation. From here on, we will lump all these sources and their categories under the title of *noise*, meaning not just unwanted sound, but anything that prevents the product or process from functioning in a smooth, seamless way. Think of noise as the friction that gets in the way of perfect performance.

When teams confront a function beset with excessive variation caused by noise, the worst possible response is to ignore the problem – the slip-it-under-the-rug response. Needless to say, this never solves the problem, although it is a surprisingly common response.

As you might expect, more proactive teams usually respond by attacking the sources of the noise, trying to buffer them, or compensating for the noise by other means. All these approaches can work to a degree, but they will almost always add to the costs.

Traditionally, companies have created new products and processes by the simple formula design-build-test, or, essentially, trial and error. This has its appeal, of course, but is ultimately time consuming, inefficient, and unimaginative. It's physically rigorous but intellectually lazy.

1.4.2 Parameter Design

Parameter design takes a different approach. Instead of using the solutions listed above, which all kick in *after* the noise is discovered, parameter design works to eliminate the effect of noise *before* it occurs by making the function immune to possible sources of variation. It's the difference between prevention and cure, the latter being one of the biggest themes of design for six sigma.

We make the function immune to noise by identifying design factors we can control and exploiting those factors to minimize or eliminate the negative effects of any possible deviations – rather like finding a natural predator for a species that's harming crops and people. Instead of battling the species directly with pesticides and the like, it's more efficient to find a natural agent. The first step toward doing this is to discard the familiar approach to quality control, which really is a focus on failure, in favor of a new approach that focuses on success.

Instead of coming up with countless ways that a system might go wrong, analyzing potential failures, and applying a countermeasure for each, in parameter design we focus on the much smaller number of ways we can make things go right! It's much faster to think that way, and much more rewarding, too. Think of it as the world of scientist versus the world of engineers. It is the goal of scientists to understand the entire universe, inside and out. A noble goal, surely, but not a very efficient one. It is the engineer's goal to understand what he needs to understand to make the product or process he's working on work well. *We need to think like engineers, looking for solutions, not like pure scientists, looking for explanations for every potential problem.*

The usual quality control systems try to determine the symptoms of poor quality, track the rate of failure in the product or process, then attempt to find out what's wrong and how to fix it. It's a backward process: beginning with failure and tracing it back to how it occurred.

In parameter design we take a different tack: one that may seem a little foreign at first, but which is ultimately much more rewarding and effective. As discussed earlier, every product or process ultimately boils down to a system whereby energy is transferred from one thing to another to create that product or process. It's how electricity becomes a cool breeze pumping out of your air conditioner. In the case of software or business processes, a system transforms information, not energy, and exactly the same optimization can be applied.

In the parameter design approach, instead of analyzing failure modes of an air-conditioning unit, we measure and optimize the variability and efficiency of the energy transformation from the socket to the cool air pumping out of the unit. In other words, we optimize the quality of energy transformation.

This forces us to define each intended function clearly so that we can reduce its variability and maximize its efficiency. In fact, that's another core issue of parameter design: the shift from focusing on what's wrong and how to fix it to focusing on what's right and how to maximize it. Mere debugging and bandaging are not effective.

To gain a deeper understanding of the distinctions between the old and new ways of thinking, it might be helpful to walk through an example. Let's look at the transfer case of a brand new four-wheel-drive truck. Now, as you probably know, the basic function of this system is as follows: The fuel system sends fuel to the engine, which turns it into active energy and sends it on to the transmission, which sends it on to the transfer case, whose job is to take that energy and distribute it to the front and rear axles for maximum traction and power. The transfer case, therefore, acts as the clearinghouse, or distribution center, for the car's energy.

Even with new transfer cases, common problems include audible noise, excessive vibration, excessive heat general, poor driving feel, premature failure or breakdown, and poor reliability. When engineers see any of these conditions, they traditionally have jumped right in to modify the transfer case's design to minimize the particular problem. The catch is, however, that often "fixing" one of these problems only makes another one worse. For example, we could reduce audible noise, only to find a dangerous increase in friction-generated heat.

It's like squeezing one end of a balloon only to see the other end expand, or quitting smoking only to see your weight increase. Using this approach, instead of eradicating the problem, we've only shifted the symptom of variability from one area to another, and have spent a lot of time, energy, and money in the process.

With parameter design, however, instead of trying to debug the transfer case bug by bug, which often results in us chasing our tails, we focus on reducing the variability of energy transformation, then minimizing the energy that goes through the transfer case cleanly. In other words, we shift our focus from defense to offense.

The theory goes like this: if we could create a perfect transfer case with zero energy loss, there would be no "wasted" energy necessary to create audible noise, heat, vibration, and so on. Sounds good, of course, but obviously building the perfect transfer case is still a pipe dream. But the thinking behind the perfect transfer case, however, can help us build a better one. Wouldn't it be better to try to achieve the perfect energy-efficient transfer case than to try to achieve perfection through endless debugging, putting out fire after fire in the hope of eliminating fires forever? As Ben Franklin said: "An ounce of prevention is worth a pound of cure." We try to build that prevention into the design. It's estimated that in a typical US company, engineers spend 80% of their time putting out fires, not preventing them. Smart companies reverse this ratio.

Usually, the single biggest source of function variation stems from how the customer uses a product or process. (Recall noise factors.) The reason is simple: Labs are sterile places where sensible scientists test the product or process under reasonable conditions, but customers can use these products in a thousand different ways and environments, adding countless variables, including aging and war. Virtually no one can anticipate the many ways that customers might be tempted to use a product or process. This is how we get warning labels on lawnmowers advising consumers not to use them on hedges.

But that's the real world. We cannot prevent customers from using their four-wheel-drive cars in just about any manner they wish. So how do we solve this problem?

Let's take a simple pair of scissors as an example. When designed well, as almost all of them are, they can cut regular paper and basic cloth well enough. But what can you do about customers who buy them to use on materials for which they were never intended, such as leather or plastic?

Most companies would do one of two things. Either they would include stern warnings in the owner's manual, and on the product itself, that the scissors are not intended for use on leather or plastic and that using them on those materials would render the warranty null and void; or companies can give up trying to educate customers, assume the worst, and bolster the design of the scissors so that they actually can cut leather and plastic.

The problem with the first approach is that such warnings only go so far; your company might still be found liable in court. In any case, even intelligent customers might be turned off by a pair of scissors that cannot cut through leather and plastic, even if they never intend to use theirs in that way. The problem with the second approach – making the scissors all but bulletproof – is that, for the vast majority of customers, the extra materials and joint strengthening would be overkill and would raise the price of the product, even for people who will never need such additional force.

With parameter design, however, you don't need to resort to either unsatisfactory solution, because the method helps you create "perfect scissors" that require virtually no effort to cut almost any material. Instead of simple bolstering the device, parameter design streamlines the product to avoid the problems that arise when the product is being used on tough materials, in much the same way that offices solved their "paper problem" not by merely building more and bigger file cabinets, but by converting their information to microfilm, microfiche, and finally to computers.

Making the scissors more efficient reduces the odds of damage and deterioration, and therefore effectively makes the scissors immune to the extremes of customer usage variation without burdening the product with undue costs.

The same concept of parameter design for robust optimization can be applied to the design of a business transactional process. Let's take the efficiency of hospital service, for example. Even for a case like this, we can look at the system as an energy transformation.

Each patient visiting an emergency room (ER) represents the input to the system. Each of them has a different level of demands. One may require a simple diagnosis and a prescription; another may require immediate surgery. The total time spent by a patient in the hospital represents the output. Therefore, we can define the ideal function as the ideal relationship between the input demanded and the actual output. Then we want to optimize the system for robustness. We want the relationship between the input and output to have the least variability at the highest efficiency.

In other words, we want the design to address the number of beds, number of nursing staff, number of health unit coordinators on staff, number of doctors on staff, pharmacy hours, in-house coverage, ER coordinator, dedicated x-ray services, private triage space, and so on. And we want the design to be the most robust against noise factors such as total number of patients visiting, time of patient visit, equipment down time, lab delay, private MD delay, absenteeism, and so on. In essence, we want the relationship between the inputs (the demands of each patient) and the outputs (the time spent on each patient) to have the smallest variability with the highest efficiently. Next, we formulate an experiment with this objective in mind which can be executed by computer simulation instead of more expensive, real-life models.

In summary, teams will learn how to apply the principles of parameter design to optimize the performance of a given system in a far more elegant fashion than just debugging or bolstering it would ever accomplish.

1.4.3   Tolerance Design

In parameter design we optimize the design for robustness by selecting design parameter values, which means defining the materials, configurations, and dimensions needed for the design. For a transfer case in a four-wheel-drive truck, for example, we define the type of gears needed, the gear material, the gear heat treatment method, the shaft diameter, and so on. For a hospital, we define the number of beds, pharmacy hours, and so on. In sum, in parameter design we define the nominal values that will determine the design.

The next step is tolerance design, in which we optimize our tolerances for maximum effect, which does not necessarily mean making them all as tight as they can be. What it does mean is making them tight where they need to be tight, and allowing looser ones where we can afford to have looser ones, thus maximizing the quality, efficiency, and thrift of our design.

For tolerance design optimization, we use the quality loss function to help us evaluate the effectiveness of changing dimensional or material tolerances. This allows us to see if our results are better or worse as we tweak a particular element up or down.

Let's start with tolerance design optimization. *Tolerancing* is a generic label often applied to any method of setting tolerances, be they tolerances for dimensions, materials, or time, in the case of a process.

Tolerance design means something more specific: a logical approach to establishing the appropriate tolerances based on their overall effect on system function (sensitivity) and what it costs to control them. As mentioned earlier, the key model employed in tolerance design is quality loss function. To say it another way, tolerance design describes a specific approach to improving tolerances by tightening up the most critical tolerances (not all of them, in other words) at the lowest possible cost through quality loss function.

This requires us first to determine which tolerances have the greatest impact on system variability, which we accomplish by designing experiments using orthogonal arrays. These experiments are done by computer simulation (occasionally, by hardware). This allows us to prioritize our tolerances – to decide which changes reap the greatest rewards – and thereby helps us make wise decisions about the status of our various options, letting us know which ones we should tighten, loosen, or leave alone.

Think of it as a baseball team's batting order, and you're the manager. Your job is to maximize run production, and you do it by trying different players in different spots in the lineup. The key is isolating who helps and who does not. Substituting various players in the lineup and changing the order will give you the results you need to determine who works best in which position.

Tolerance design will help teams meet one of the primary objectives of the initiative: developing a product or process with six sigma quality while keeping costs to a minimum. Tolerance design is intended to help you and your team work through the process of establishing optimal tolerances for optimal effect.

The goal is not simply to tighten every standard, but to make more sophisticated decisions about tolerances. To clarify what we mean, let's consider a sports analogy. Billy Martin was a good baseball player and a great manager. He had his own off-field problems, but as a field general, he had no equal. One of the reasons he was so good was because he was smart enough, first, to see what kind of team he had, then to find a way to win with them, playing to their strengths and covering their weaknesses – unlike most coaches, who have only one approach that sometimes doesn't mesh with their players.

In the 1970s, when he was managing the Detroit Tigers, a big, slow team, he emphasized power: extra base hits and home runs. When he coached the Oakland A's a decade later, however, he realized that the team could never match Detroit's home-run power, but they were fast, so he switched his emphasis from big hits to base stealing, bunting, and hitting singles. In both places, he won division crowns, but with very different teams.

It's the same with tolerance design. We do not impose on the product or process what we think should happen. We look at what we have, surmise what improvements will obtain the best results, and test our theories. In Detroit, Martin didn't bother trying to make his team faster and to steal more bases because it wouldn't have worked. He made them focus on hitting even more home runs, and they did. In Oakland, he didn't make them lift weights and try to hit more homers, because they didn't have that ability. He made them get leaner and meander and faster and steal even more bases. And that's why it worked: he played to his team's strengths.

You don't want to spend any money at all to upgrade low-contributing tolerances. You want to reduce cost by taking advantage of these tolerances. You don't want to upgrade a high-contributing tolerance if it is too expensive. If the price is right, you will upgrade those high contributors. Tolerance design is all about balancing cost against performance and quality.

One common problem is that people skip parameter design and conduct tolerance design. Be aware of the opportunities you are missing if you skip parameter design. By skipping parameter design, you are missing great opportunities for cost reduction. You may be getting the best possible performance by optimizing for robustness, but if the best is far better than required, there are plenty of opportunities to reduce cost. Further, you are missing the opportunity to find a bad concept, so that you can reject the bad concept at the early stage of product/process development. If the best concept you can come up with is not good enough, you have to change the concept.

The result of tolerance design on designs that have not been optimized is far different from the result of tolerance design after robust optimization has taken place. In other words, you end up tightening tolerances, which would have been unnecessary if the design had been optimized for robustness in the first place. Think of all firefighting activities your company is doing today. If the design were optimized, you would have fewer problems and the problems would be different. Hence, solutions would be different.

2

Eight Steps for Robust Optimization and Robust Assessment

We will take you through Eight Steps of Robust Optimization then Eight Steps of Robust Assessment. The eight steps are shown in Figure 2.1.

Robust Assessment is a test procedure designed to benchmark "Robustness" of multiple design concepts. Typical purposes of Robust Assessment are:

1. benchmarking competitors' products so that we can make an objective judgment on how good our products are and also to set strategic target for our next product;
2. evaluating multiple design alternatives so that we can make a correct decision on which design concept to develop further;
3. comparing multiple suppliers' design solutions.

On the other hand, Robust Optimization is a test procedure to optimize "Robustness" of a particular design concept under development. The following describes Robust Optimization.

- Given a design concept, it will attempt to achieve perfection of the finished product.

Robust Optimization: World's Best Practices for Developing Winning Vehicles, First Edition. Subir Chowdhury and Shin Taguchi.
© 2016 Subir Chowdhury, Shin Taguchi, and ASI Consulting Group, LLC.
Published 2016 by John Wiley & Sons, Ltd.

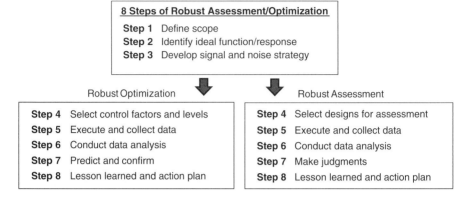

Figure 2.1   Eight steps for Robust Optimization and Robust Assessment

- It strives to achieve the most robust combination of design parameters, also called control factors, such that the design is the most robust at low cost.
- It optimizes "the function" of product/process for robustness.
- It uses "Robustness of function" as the criterion for optimization, as opposed to "Meeting requirements" as the criterion.
- Robust Assessment is a subset of Robust Optimization as the first three steps are the same.

2.1   Before Eight Steps: Select Project Area

It is strongly recommended that a project area be selected by catch-balling between executives, R&D and engineering management. For instance, new technology such as "Stop and Start" can be identified as a project area. A project area is sometimes called a mega-project or theme. Then several optimization scopes can be strategically identified.

There have been so many situations where the engineering team end up with asking for help after spending the majority of the planned development time. It is human nature that people ask for help only when they find out they are in trouble. Let's not wait for a problem to take place. As the previous chapter emphasized, "Robust Optimization" does not equal "Meeting Specs and Requirements." Optimization for robustness leads to the following benefits:

- Optimization is to seek for the best the design can be.
- There is no guarantee that the optimized design will meet all requirements. The optimized design has the highest probability to meet all requirements.

- The optimized design may become much better than just meeting all requirements. When that is the case, there will be plenty of opportunities for cost and weight reduction.
- The optimized design may not meet some requirements. Or it may not meet any requirement. It all depends on how good the design concept is. You cannot achieve a robust design by optimizing a lousy design concept.
- Finding the lousy design concept at an early stage in development is extremely critical for management of product development. Do you want to spend resources to develop a lousy design?

Once a design concept is identified for development, why not optimize its function for "Robustness"? Let's not wait for validation team checks for "Robustness." Recognize the difference between "Optimizing for Robustness" and "Checking Robustness."

Another consideration is to address a cross-functional mega-theme such as Fuel Economy and Quiet Interior. Then the cross-functional engineering team must identify several optimization projects. Each optimization project will go under Eight Steps of Robust Optimization.

2.2 Eight Steps for Robust Optimization

2.2.1 Step 1: Define Scope for Robust Optimization

The scope for each optimization project can be small or large. Optimization of fastening function with a set of good old nuts and bolts seems to have very small scope, but it can be critical to assure robustness for the sake of safety.

When a system is large, it is usually practical to conduct Robust Optimization at subsystem levels. For instance, a sunroof system can be deconstructed into three critical functions: (1) sun roof displacement function; (2) closing and sealing function to prevent water leakage and wind noise when it is closed; and (3) air flow function when it is open to prevent excessive wind noise.

The scope can be huge as long as it is measurable. For instance, one case study in this book optimized an angular displacement sensing system. This is a new measurement system consisting of three subsystems, namely, (1) sensor for detection; (b) IC for conversion; and (c) software for computations. All three subsystems must work in harmony to perform robust sensing function. Traditionally, three subsystems are developed separately to meet each set of requirements, and then the whole system is validated. The study had optimized 48 design control factors from three subsystems simultaneously. They claimed the time to market was reduced to one third.

Table 2.1   Considerations for scoping optimization/assessment project

- ❖ Be strategic and preventive
- ❖ Consider mega project
- ❖ Consider cross-functional mega project
- ❖ Prioritize based on criticality of a function
- ❖ Availability of measurement technology
- ❖ Feasibility of computer simulations

This sensing system optimization with all three subsystems was possible because they were able to develop an Excel based simulation-engine to simulate all three functions simultaneously. In other words, they developed a means to measure the system response as a function of parameters of all three subsystems. It is critical for engineering management to seek measurement technology that enables optimization with a big scope. Time to market is becoming more and more competitive. Table 2.1 lists considerations for scoping.

2.2.2 Step 2: Identify Ideal Function/Response

2.2.2.1 Ideal Function: Dynamic Response

After the scope is defined, the next is to identify the Ideal Function. The Ideal Function can be recognized by "Energy Thinking."

Energy thinking is nothing but thinking of basic physics. For instance, let us look at the opening and closing of automotive windows. As a user turns on the switch to open/close the window, the motor draws electrical current and moves the window up and down. Therefore the input to this system is the electrical energy. The system converts input energy to output energy. The output is the movement of the window. This output can be called the work done by this system. Don't we want to optimize our work? Figure 2.2 shows input and output based on energy thinking.

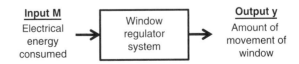

Figure 2.2   Input and output based on energy thinking

It is ideal to transform 100% of input energy into the intended output energy/work. "Inefficiency" and "Variability" of this energy transformation result in numerous failure modes and quality problems, such as slow movement, uneven movement, squeak noise, vibrations, no movement and short life. Poor quality and poor reliability are nothing but symptoms of poor energy transformation.

The ideal function is the ideal relationship between the input and the output based on energy thinking. The input is also called "Input Signal" and is denoted by M. The output is called "Output Response," denoted by y. Table 2.2 shows some examples from real case studies.

1. The most common form of ideal function is $y = \beta M$. This means when the input is zero then the output is zero, and the output is proportional to the input energy. In general, the smaller the variability of this relationship, the higher the slope β is desired.
2. Input and output do not have to be exactly energy. They can be "Power," which is simply the rate of energy or energy per unit time. It can be some measure that represents energy/power.

   Or, it can be something that is not energy. For instance, $M =$ True Value and $y =$ Measured Value are typically used for optimizing a measurement system. Any measurement system uses some kind of energy transformation and the results are converted into measured values. In any case, you need to use energy thinking to identify the ideal function.
3. The system does not have to be hardware. Software's function is to transform "Information." The ideal function is recognized by focusing on "Work" done by the software. An autonomous driving system has various functions that are performed by both the hardware and software.
4. Ideal Function is referred to as a "Dynamic Response." It is called dynamic because the response is not just a simple scalar. What is being optimized is a function which is defined as the relationship between the input and output.

2.2.2.2 Nondynamic Responses

Use of Dynamic Response is the most effective for Robust Optimization/ Assessment. Therefore the development team must have sufficient discussion on the ideal function. In some cases, the ideal function cannot be defined due to lack of technical competence and creativity. Other times, it is defined but it

Table 2.2 Various ideal functions

System	Objective	Input signal M	Response y	Ideal function
Ski	Turn	Weight Shift	Amount of Turn	$y = \beta M$
Air Pump	Increase Pressure	Force x Displacement	Δ Pressure	Exponential
Bathroom Scale	Measure Weight	True Weight	Measured Weight	$y = \beta M$
Golf Shot	Carry the ball	Head Velocity ^ 2	Distance	$y = \beta M$
Emergency Room	Take care of patients	Total Man-Hour	Weighted Patient Need	$y = \beta M$
True-False Test	Evaluate Student	Correct Solution	Answer	$0 \rightarrow 0 \ \& 1 \rightarrow 1$
Investment	Make Money	Time	Growth	Exponential
Prediction System	Predict	True Value	Predicted Value	$y = \beta M$
Pedestrian Protection	Predict Time to Impact	True Value	Predicted Value	$y = \beta M$
Wind Mill Generator	Produce Power	Air Flow x Pressure	Electric Power	$y = \beta M$
HVAC Fan	Generate Airflow	Voltage	Air Flow Rate	$y = \beta M$
HVAC Fan	Adjust Airflow	Voltage Setting	Air Flow Rate	$y = \beta M$
Heating	Generate Heat	Consumed Energy	Δ Temp	Exponential
Cooling	Reduce Temperature	Consumed Energy	Δ Temp	Exponential
Control System	Adjust Parameters	Demanded Compensation	Actual Compensation	$y = \beta M$
Automotive Brake	Reduce Speed	Force on Pedal	Fluid Pressure	$y = \beta M$
Automotive Brake	Reduce Speed	Fluid Pressure	Decel	$y = \beta M$
Steering	Turn	Wheel Displacement	Lateral Acceleration	$y = \beta M$
Lighting	Produce Light	Consumed Energy	Luminance x Time	$y = \beta M$
Automobile Impact	Protect Cabin	Impact Energy	Energy Obsorved	$y = \beta M$
Transmission	Transmit Power	Input Power	Output Power	$y = \beta M$
Transmission	Transmit Power	Input Power	Output Power	$y = \beta M$
Power Sliding Door	Open/ Close Door	Consumed Energy	Door Displacement	Ideal Profile
Switch Feel	Provide Good Feel	Force	Displacement	$y = \beta M$
Machining	Remove Material	Consumed Energy	Amount Removed	Ideal Profile
Chemical Reaction	$A + B \rightarrow C + D$	Time	Fraction Reacted	Exponential

cannot be used due to lack of measurement technology or high cost of measurement. In those cases, the team will settle for a nondynamic response. Table 2.3 summarizes nondynamic responses.

When the ideal function/response is identified, it is a good time to start developing the P-diagram (see Figure 2.3). Input, Output, Control Factors, Noise Factors and Failure Modes are five elements in the P-diagram. "P" stands for parameters. Noise factors will be discussed in Step 3 and Control Factors in Step 4.

2.2.3 Step 3: Develop Signal and Noise Strategies

In case of ideal function (as opposed to nondynamic response), the output response will be measured as the input signal and noise factors are varied intentionally.

2.2.3.1 How Input M is Varied to Benchmark "Robustness"

It is good to have a policy that says we assure performance at all customers' usage conditions. With that said, the range of input must cover all real customer usages. The input must be varied at least three levels, say "Low," "Med" and "High."

For a brake subsystem in the P-diagram (Figure 2.4), the input is the brake fluid pressure and it ranges from 0 to 2000 psi, or 0.0 to 14.0 Mpa from no braking to hard braking. Therefore the fluid pressure must be varied low to high to cover the entire range. Sometimes the range may be expanded wider than customer usage so the limitation of the design can be checked. Note that in some cases, the input is also a measured value.

2.2.3.2 How Noise Factors Are Varied to Benchmark "Robustness"

Noise factors are all those causes of variability of energy transformation (Figure 2.5). Table 2.4 summarizes five categories of noise factors.

Table 2.5 shows five types of countermeasures that can be applied to deal with noise factors. Robust Optimization is the countermeasure type IV, which is used to minimize the influence of all noise factors.

In order to assess robustness, noise factors are varied intentionally. Table 2.6 lists strategies to vary noise factors. The flow chart (Figure 2.6) shows the thought process to identify the most effective noise strategy. The

Table 2.3 Nondynamic responses

Nondynamic Response	Description	Examples	
Nominal-the-Best Type-1	Response with a target value. This response typically relates to energy/work. It does not take a minus value.	Door Closing Effort	System Voltage
		Head Lamp Brightness	Window Displacement Rate
Nominal-the-Best Type-2	Response with a target value. This response does not related to energy/work. It can take a minus value.	Tire Alignment	Steering Wheel Alignment
		Gap between Door and Body	Headlamp Aim
Smaller-the-Better	Zero is the ideal value. This response does not take a minus value.	Noise and Vibration at Seat	Fuel Leakage
		Emission	Failure Rate
Larger-the-Better	Infinity is the ideal value. It should be recognized as Type-1Nominal-the-Best, since in most cases it is not practical to assume it becomes infinity.	Pressure to start leak	Structure Stiffness
		Operating Window	Time to failure
Classified Attribute	Rating of Goodness and Badness Typically a subjective rating	Seat Comfort Rating	Consumer Report Rating
		JD Power Score	Switch Operation Feel
Operating Window	Given a design parameter, the distance between failure due to not enough energy and failure due to too much energy	For cancer treatment, dosage to kill cancer cells and Dosage to kill healthy cells	
		For a paper feeding mechanism, Spring Force to Mis-feed and Spring Force to Multi-feed	

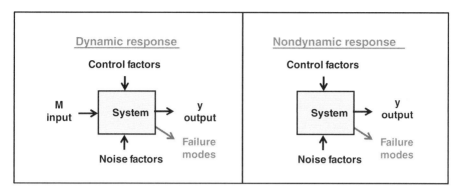

Figure 2.3   P-diagrams

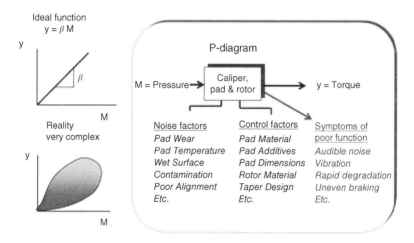

Figure 2.4   P-diagram on automotive brake subsystem

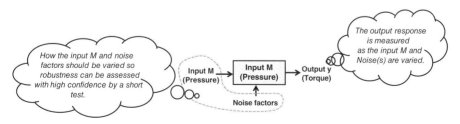

Figure 2.5   Strategy for how signal and noise should be varied

Table 2.4 Category of noise factors with example

	Noise factor category	Description	Examples
1	Environment	Environment during usage. Customer condition.	Temperature, humidity | Ultra violet, water intrusion Altitude, road condition | Vibrations, loading
2	Aging	Wear, deterioration, change of material property over time	Wear and tear | Vibration cycling Temperature cycling | Loading cycling
3	Manufacturing variation	Mfg and assembly variation, piece to piece and lot to lot variation	Variation within tolerances | Misalignments after assembly Material variation | Tolerance stack-up
4	Surrounding subsystems	Status of surrounding subsystems	For a steering system, what suspension is doing | For a braking system, what steering is doing
5	Misuse/evil intent	Misuse and abuse, terrorism act, destructive action	Use of latch as can openers | Terrorist attack on nuclear power plant Act of automobile theft

Table 2.5 Countermeasures for noise factors

	Countermeasures for noise factors	Examples	Comments
I	Ignore them	Use of champion data only. Then test to see meeting requirements.	Ignoring critical noises will cause disasters in later stages such as high warranty and fire fighting
II	Control or eliminate them	Inspections, process control, quality, assurance activities, fool proofing, tighten tolerances	It is cost-performance trade off. If cheap, control. But we can not control customers usage and aging effect.
III	Compensate for their effects	Voltage regulation circuit, ABS for brake, engine control, subsystems for autonomous driving	Feedback control and feedforward (adaptive) control are to do this. Optimize compensation system for robustness!
VI	Desensitize or minimize their effects	Robust optimization (also called parameter design)	Explore design parameters or control factors to maximize robustness at the lowest cost
V	Innovate robust design concept	Transistor to replace vacuum tube Modularized automotive interior design	Create and select a new design concept that is more robust than any previous design concepts

Table 2.6 Some popular noise strategies for robust assessment with examples

	Noise Strategy	Description	Example
a	One dominant noise	Use just one noise factor. When robustness against this dominant noise can assure robustness to all other noises	T: Temperature as noise: T1 = -20C T2 = 20C T3 = 80C
b	Compounded noises	By compounding a few to several noise factors to generate two noise conditions N1 and N2, such that N1=Low Response and N2=High Response	N: Compounded Noise N1= Fuel Flow is low = High RVP, Low Temp, Low Tank Pressure N2= Fuel Flow is High= Low RVP, High Temp, High Tank Pressure
c	Full factorial	All possible combination of a selected noise factors	W: Aging W1= New and W2= After Aging T: Temp T1= Low and T2= High Four Noise Conditions: W1T1 W1T2 W2T1 W2T2
d	Use of orthogonal array	Assign multiple noise factors to an orthogonal array	When simulating, take variability of all control factors
e	Hybrid of a, b and c	Mixture of a, b, and c above. Compounding many noises blinded can be dangerous	W: Aging W1=New and W2-Aged N: Compounded Noise: N1=Low Response N2= High Response Four Noise Conditions: W1N1 W1N2 W2N1 W2N2

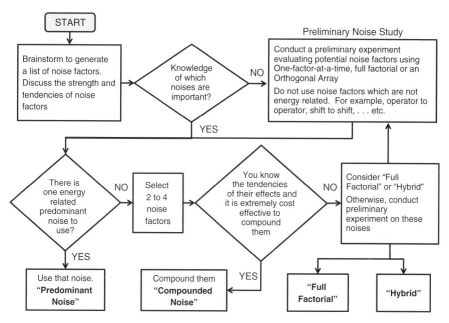

Figure 2.6   Flow chart for developing effective noise strategy

challenge is to develop a strategy to vary noise factors to satisfy the following criteria:

1. Effectiveness to assess robustness: The ideal state is "If the design is robust against selected noise strategy, it will be robust against all other noises." Such noise strategy can be standardized and used for family and future products.
2. Cost effectiveness to run test/simulation: Develop noise strategy that can be applied in short time at low cost.

It takes creative thinking to develop a great noise strategy.

For brake subsystem optimization, the input signal and noise factors are varied as shown in Figure 2.7. It was a known fact that N1 tends to be lower torque than N2 and S1 tends to be much lower torque than S2.

What we have done is to develop a test for assessing "Robustness" of torque generation function of this subsystem. Robust Assessment for benchmarking is to measure this set of data from competitors' designs.

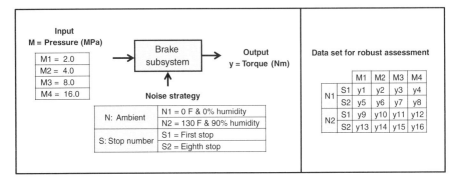

Figure 2.7   Data set for assessing "robustness"

We have just completed the formulation of robust assessment. It is now critical to get a feel for confidence in our formulation. Is the ideal function behaving as it should? Is the noise strategy generating variability as we expected? Is the noise strategy effective such that measurement error is smaller than the noise effect? Can we trust the measurements? Is this an effective way to assess "Robustness"?

In order to answer these questions, it is strongly recommended to obtain data to compare a few designs. For instance, let the following be the dataset for three designs, namely A1 = Our Current Product, A2 = Target Competitor's Product and A3 = Best in Class Product. The table and graphs in Figure 2.8 show the result.

The ideal function $y = \beta M$ is based on physics and it implies:

- It is based on physics that the output braking torque y is proportional to the input fluid pressure M.
- A smaller variability in relationship between M and y is more robust and more desirable.
- More linearity between M and y is desirable.
- A higher slope between M and y is desired for the sake of "Efficiency of Energy Transformation."

Let's look at the graphs and see what the data is telling us. Look at the data plots and characterize the difference between the three designs.

- A1 has more variability than A2 and A3.
- A2 and A3 not only show smaller variability, they show the torque increasing more linearly than A1.

| | M1 = 2.0 | | | | M2 = 4.0 | | | | M3 = 8.0 | | | | M4 = 16.0 | | | |
| | N1 | | N2 | | N1 | | N2 | | N1 | | N2 | | N1 | | N2 | |
	S1	S2	S1	S2	S1	S2	S1	S2	S1	S2	S1	S2	S1	S2	S1	S2
A1: Our Current	110	320	165	395	180	430	320	500	860	1170	920	1215	1850	2210	1865	2445
A2: Target Competitor	200	265	230	285	450	595	510	620	870	1065	920	1005	1815	2020	1915	2115
A3: Best in Class	265	300	275	350	630	675	680	725	1185	1280	1290	1375	2395	2425	2410	2500

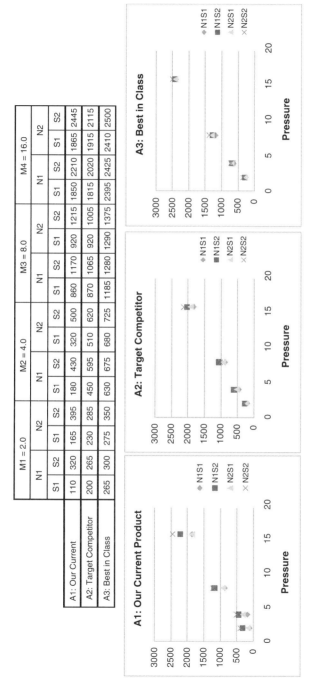

Figure 2.8 Data set and graphs for checking validity of formulations

- A3 is producing more torque than A2. Maximum torque of A2 is around 2000 Nm and that of A3 is 2500 Nm. Therefore A3 is more desirable than A2. A3 is the best, then A2 and A1.
- We can conclude that our formulation of Robust Assessment, namely, the ideal function and noise strategy, is effective and it is discriminating good design from poor design in terms of robustness.
- Noise effects are an overwhelming measurement error since noise factors have consistent tendencies.

As we see here, studying graphs can be sufficient to make judgments. It is always recommended to plot raw data and study them. However, Signal to Noise Ratio provides much better resolutions. The computations of S/N will be demonstrated in Step 6.

2.2.4 Step 4: Select Control Factors and Levels

2.2.4.1 Traditional Approach to Explore Control Factors

The typical product development process is shown in Figure 2.9. Prototypes are developed and tested by simulation or by hardware. Various tests are done to check if the design meets numerous requirements for performance, quality, reliability, and durability. The question is "Does the design meet requirements?" When some requirement is not met, the design is studied and modified by tweaking design parameters and is then tested again. This "Design-Build-Test-Fix" cycle is repeated until all requirements are met. In some cases, it will appear as a cruel endless cycle. We refer to this approach as a Whack-a-Mole Development Process.

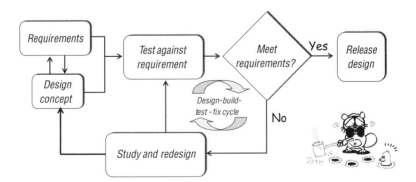

Figure 2.9 Traditional product development

2.2.4.2 Exploration of Design Space by Orthogonal Array

The assessment criterion for Robust Optimization is not "Whether the design meets requirements." It is "Robustness" of what the design is intended to do. Robust Optimization is to seek for the combination of design parameters such that robustness is maximized.

Design parameters are things engineers can specify, such as choices of Material, Dimensions, Spring type, Spring constant, Shape of cross-section, Resistance value, Capacitance value, Constants in software, Entries to look up table, Amount of chemical agent, Type of additive, etc. Because they are controllable by design engineers, they are called "Control Factors." Table 2.7 shows example of control factors and levels for optimization of brake subsystem.

Control factor A is "Pad Material." "A1 = Type 1" and "A2 = Type 2" are called levels of factor A, which are choices for control factor A. The question now is which level is better A1 or A2, and by how much? Similarly, B, C, . . . , are control factors and each of them happens to have three alternatives or three levels. Lists of control factors and levels are generated by brainstorming. Determinations are done by considering various criteria such as Cost reduction opportunity, Weight reduction opportunity, Design for X requirements, etc.

2.2.4.3 Try to Avoid Strong Interactions between Control Factors

Another critical consideration is to avoid strong interactions among control factors. Let's take donut production as an example, as shown in Figure 2.10.

Table 2.7 Control factors and levels

Control Factor	Level-1	Level-2	Level-3
A: Pad Material	Type-1*	Type-2	
B: Pad Shape e/w	4	5*	6
C: Pad Curve Profile	Type-1	Type-2*	Type-3
D: Pad Additive	Low	Med*	High
E: Rotor Material	Gray	Cast*	Steel
F: Pad Taper	Low	Med*	High
G: Tapering Thickness	Low	Med*	High
H: Rotor Structure	Type-1	Type-2*	Type-3

* Indicates the baseline design level.

The current donut production process results in too many undercooked donuts and only 66% are good donuts. So an experiment was executed. Both increasing "Oil Temperature" and "Time" made an improvement when applied separately. The question now is what happens when both temperature and time are increased (Figure 2.11)?

The answer is "Only God knows." It may become 100% because just the right amount of heat is applied. It may become 0% because all of them are overcooked.

An interaction is defined as "The effect of one factor on the output is different depending on the value of other factor." In this case, increasing temperature can decrease or increase the percentage of Good Donuts depending on frying time.

We must avoid this type of unpredictable strong interaction between control factors. Otherwise, development activities become cumbersome and inefficient because of numerous strong interactions. Unfortunately, this is what we see across industries.

Having a high percentage of good donuts is just a requirement. Resulting in a low percentage of good donuts is just a symptom of variability. So what is an

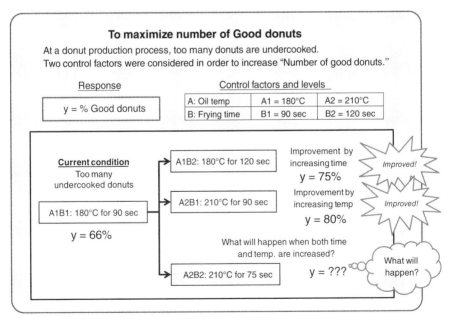

Figure 2.10   Maximization of number of good donuts

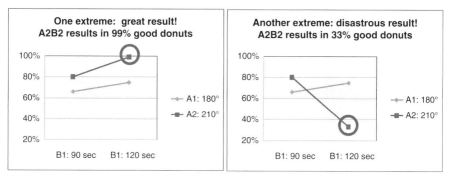

Figure 2.11   What happens when both temperature and time are increased?

effective approach in order to increase "percentage of Good Donuts"? We should measure the function which is "Heat Penetration" and reduce variability of "Heat penetration." This is a much stronger strategy as shown in Figure 2.12.

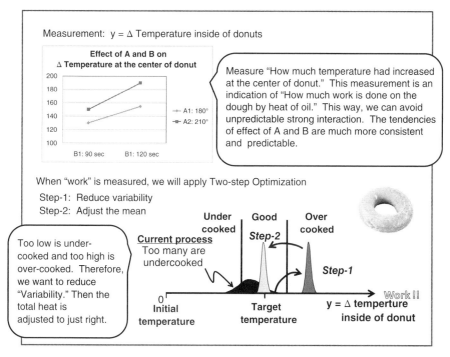

Figure 2.12   Energy response and two-step optimization

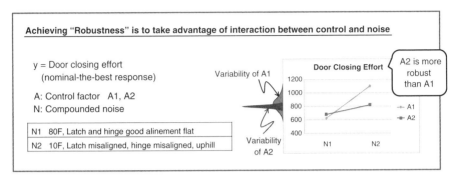

Figure 2.13 "Robustness"= "interaction between control factor and noise factor

By the way, achieving "Robustness" is to take advantage of weak interaction between Control factor and Noise factor (Figure 2.13).

So as shown in the Figures, the response y = Increase in temperature of donut dough as a Nominal-the-Best response and Two-step optimization is applied. This is what we mean by "Energy Thinking".

2.2.4.4 Orthogonal Array and its Mechanics

Orthogonal arrays provide a powerful method for exploring control factor combinations. Table 2.8 is the standard L_{18} orthogonal array and 18 recipes generated by L18.

Recipe No. 1 is the first row, which is A1, B1, C1, D1, E1, F1, G1 and H1. Recipe No. 1 is to use A1 = Type-1 Pad Material, B1 = Pad Shape e/w of 4, C1 = Type-1 Pad Curve Profile, D1 = Low Pad Additive, etc. This way, 18 recipes are specified by L18.

The term "orthogonal" means "balanced" or "separable." Let's look at the columns for A and B. Within A1 experiments, B1 occurs three times, B2 three times and B3 three times. This pattern is called "balance" or "orthogonality." Similarly, each of G1, G2 and G3 has H1, H2 and H3 replicated three times, and vice versa. Any two factors are orthogonal to each other.

Orthogonality may not sound familiar, but the notion has been around since some time before the year 1 AD. Without knowing it, you are very familiar with orthogonality in your everyday life. Take football games, for example. In the first half, one team kicks off and in the second half, the other team kicks off. Every quarter they exchange ends of the field. Why do they do that? In a tournament, half the games are played at home and the other half away. These are done such that comparisons are fair.

Table 2.8 Orthogonal array L18 and experimentation recipes

Standard L18
$L_{18}(2^1 \times 3^7)$

L_{18}	A (1)	B (2)	C (3)	D (4)	E (5)	F (6)	G (7)	H (8)
1	1	1	1	1	1	1	1	1
2	1	1	2	2	2	2	2	2
3	1	1	3	3	3	3	3	3
4	1	2	1	1	2	2	3	3
5	1	2	2	2	3	3	1	1
6	1	2	3	3	1	1	2	2
7	1	3	1	2	1	3	2	3
8	1	3	2	3	2	1	3	1
9	1	3	3	1	3	2	1	2
10	2	1	1	3	3	2	2	1
11	2	1	2	1	1	3	3	2
12	2	1	3	2	2	1	1	3
13	2	2	1	2	3	1	3	2
14	2	2	2	3	1	2	1	3
15	2	2	3	1	2	3	2	1
16	2	3	1	3	2	3	1	2
17	2	3	2	1	3	1	2	3
18	2	3	3	2	1	2	3	1

18 recipes specified by L18

	A: Pad Material (1)	B: Pad Shape e/w (2)	C: Pad Curve Profile (3)	D: Pad Additive (4)	E: Rotor Material (5)	F: Pad Taper (6)	G: Tapering Thickness (7)	H: Rotor Structure (8)
1	Type-1	4	Type-1	Low	Gray	Low	Low	Type-1
2	Type-1	4	Type-2	Med	Cast	Med	Med	Type-2
3	Type-1	4	Type-3	High	Steel	High	High	Type-3
4	Type-1	5	Type-1	Low	Cast	Med	High	Type-3
5	Type-1	5	Type-2	Med	Steel	High	Low	Type-1
6	Type-1	5	Type-3	High	Gray	Low	Med	Type-2
7	Type-1	6	Type-1	Med	Gray	High	Med	Type-3
8	Type-1	6	Type-2	High	Cast	Low	High	Type-1
9	Type-1	6	Type-3	Low	Steel	Med	Low	Type-2
10	Type-2	4	Type-1	High	Steel	Med	Med	Type-1
11	Type-2	4	Type-2	Low	Gray	High	High	Type-2
12	Type-2	4	Type-3	Med	Cast	Low	Low	Type-3
13	Type-2	5	Type-1	Med	Steel	Low	High	Type-2
14	Type-2	5	Type-2	High	Gray	Med	Low	Type-3
15	Type-2	5	Type-3	Low	Cast	High	Med	Type-1
16	Type-2	6	Type-1	High	Cast	High	Low	Type-2
17	Type-2	6	Type-2	Low	Steel	Low	Med	Type-3
18	Type-2	6	Type-3	Med	Gray	Med	High	Type-1

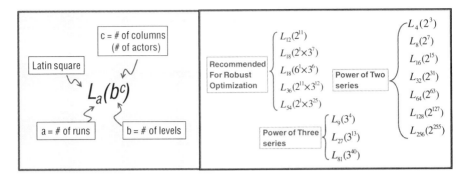

Figure 2.14 Popular orthogonal arrays

Depending on the number of control factors and number of levels, the appropriate orthogonal array is selected. Figure 2.14 shows choices of orthogonal arrays. Various useful techniques exist to modify an orthogonal array to fit your required number of factors and levels.

We are now ready to marry Step 3 and Step 4. Step 3 has specified how the data set looks for Robust Assessment. This data array is called "Outer Array." Step 4 has specified recipes of design parameters using an orthogonal array. This array is called "Inner Array." We are ready to conduct robust assessment (Outer Array) for each of 18 recipes (Inner Array). The inner array is owned by development engineers and the outer array is what the customer experiences (Figure 2.15).

2.2.5 Step 5: Execute and Collect Data

Develop a master plan for executing test and data collection. Who, when, how, and what needs to be done. Develop a plan for making samples, running computer simulation, data acquisition, etc. Make a plan for whatever it takes to produce good data. Make sure things go right as planned. Once all data collections are completed, we have what Table 2.9 shows.

2.2.6 Step 6: Conduct Data Analysis

It is not in the scope of this book to discuss details of data analyses. Because Signal-to-Noise Ratio (S/N) is very unique and an effective measure for assessing "Robustness," you need to get the idea of it and understand what it means. Understand how it can improve your product development.

Signal-to-Noise Ratio (S/N) and the slope β are computed from each of robust assessment data set (Figure 2.16). Let us go over the computation and interpretation of S/N for $y = \beta M$ as the ideal function.

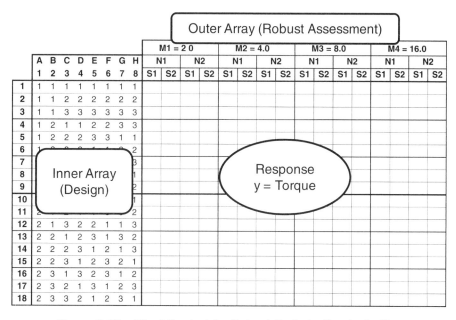

Figure 2.15 What the test for Robust Optimization looks like

2.2.6.1 Computations of S/N and β

Equations are shown in Figure 2.17 and Figure 2.18. Then the data of No.1 of L18 is used to demonstrate computations.

Figure 2.19 shows what this decomposition: $S_T = S_\beta + S_{Noise}$ is all about. Just like the Pythagorean Theorem, the total sum of squares of all data is decomposed into two components. Each component has a simple technical meaning. In general, S_β relates to the efficiency of energy transformation, therefore we want to maximize it. And we want to minimize S_{Noise} as it is variability due to noises.

- β is the slope of the best fit line through "Zero point and all data points." The higher the slope β, the larger the S_β is. The higher the slope β is, the more braking torque is generated at a given pressure. Higher β or larger S_β implies higher efficiency of energy transformation, which is usually a good thing. If it's too high, it will be adjusted to just right "Sensitivity" after maximizing S/N. Such adjustment is "Step 2 of two-step optimization."
- The larger the S_{Noise} is, the larger the variability of braking torque. The larger S_{Noise} is, the stronger the effects of noises. The larger S_{Noise} is, the larger the variance due to noise $V_{Noise} = \sigma^2_{Noise}$, which is a bad thing.

Table 2.9　Result of execution of experiment and data collection

	A	B	C	D	E	F	G	H	M1= 2.0 N1 S1	N1 S2	N2 S1	N2 S2	M2= 4.0 N1 S1	N1 S2	N2 S1	N2 S2	M3= 8.0 N1 S1	N1 S2	N2 S1	N2 S2	M4= 16.0 N1 S1	N1 S2	N2 S1	N2 S2
	1	2	3	4	5	6	7	8																
1	1	1	1	1	1	1	1	1	45	240	40	290	325	425	340	575	660	1020	810	1250	1635	1845	1725	2175
2	1	1	2	2	2	2	2	2	125	225	160	285	480	625	500	650	1015	1175	1070	1255	1800	2100	1805	2160
3	1	1	3	3	3	3	3	3	260	295	295	340	465	530	510	570	1100	1175	1125	1215	2015	2145	2030	2190
4	1	2	1	1	2	2	3	3	105	225	150	285	445	605	525	715	845	1105	1000	1210	1700	2050	1880	2120
5	1	2	2	2	3	3	1	1	105	325	160	390	345	615	430	660	860	1165	915	1215	1845	2215	1860	2445
6	1	2	3	3	1	1	2	2	210	250	215	290	470	575	495	615	840	1040	925	1050	2010	2150	2050	2155
7	1	3	1	2	1	3	2	3	200	260	225	280	455	590	505	615	875	1060	915	1000	1810	2015	1910	2110
8	1	3	2	3	2	1	3	1	0	120	140	215	200	335	180	360	555	815	615	915	1390	1505	1530	1715
9	1	3	3	1	3	2	1	2	240	315	305	390	465	605	595	675	980	1220	1115	1315	2015	2425	2200	2510
10	2	1	1	3	3	2	2	1	0	105	0	145	0	245	210	370	475	915	540	885	1315	1600	1405	1765
11	2	1	2	1	1	3	3	2	60	245	90	380	325	565	340	765	750	1170	860	1255	1660	2005	1775	2525
12	2	1	3	2	2	1	1	3	220	255	220	320	390	505	425	560	935	1085	1005	1105	2060	2155	2075	2220
13	2	2	1	2	3	1	3	2	0	105	30	270	61	335	115	365	470	670	555	820	1395	1945	1555	2165
14	2	2	2	3	1	2	1	3	250	295	260	340	600	665	665	710	1155	1245	1270	1315	2315	2395	2360	2485
15	2	2	3	1	2	3	2	1	0	160	90	195	160	435	255	480	395	660	555	975	1605	1910	1650	2125
16	2	3	1	3	2	3	1	2	135	205	220	295	435	615	460	685	945	1215	1010	1275	1950	2215	2120	2385
17	2	3	2	1	3	1	2	3	40	115	105	160	400	510	440	625	825	1080	1020	1180	1620	1940	1830	2055
18	2	3	3	2	1	2	3	1	0	60	60	255	115	390	250	650	555	1015	620	1060	1580	2005	1600	2255

A1	M1	M2	M3	M4
	2	4	8	16
N1S1	y1	y2	y3	y4
N1S2	y5	y6	y7	y8
N2S1	y9	y10	y11	y12
N2S2	y13	y14	y15	y16

A1	M1	M2	M3	M4	Linear tendency
	2	4	8	16	
N1S1	y1	y2	y3	y4	L1
N1S2	y5	y6	y7	y8	L2
N2S1	y9	y10	y11	y12	L3
N2S2	y13	y14	y15	y16	L4

Linear tendency is the sum of product of values of M and y
The larger the value of L, the higher the slope is.

Figure 2.16   Data set for Robust Assessment

Step 1 of 4: Pretreatment of data

Compute linear tendency L for each noise condition.

$$L_1 = M_1 y_1 + M_2 y_2 + M_3 y_3 + M_4 y_4$$
$$L_2 = M_1 y_5 + M_2 y_6 + M_3 y_7 + M_4 y_8$$
$$L_3 = M_1 y_9 + M_2 y_{10} + M_3 y_{11} + M_4 y_{12}$$
$$L_4 = M_1 y_{13} + M_2 y_{14} + M_3 y_{15} + M_4 y_{16}$$

Compute r = sum of squares of values of M for all data.

$$r = 4 \times \left(M_1^2 + M_2^2 + M_3^2 + M_4^2 \right)$$

Sum of squares of values of M for all data. This is the sum of squares of M1, M2, M3, M4 multiplied by 4.

Compute the sum of all linear tendencies denoted by ΣMY

$$\Sigma MY = \sum_{i=1}^{4} L_i = L_1 + L_2 + L_3 + L_4$$

ΣMY is the sum of product of all M and y pairs.

Figure 2.17   Step 1 of 4 for computations of S/N and β

Computations for S/N and β are completed by the steps shown in Figure 2.20 and Figure 2.21. The concept of S/N ratio and β is shown in Figure 2.22.

- S/N is in decibels and the higher the S/N, the more the robustness.
- "Log" is taken for better "Additivity," or better ability to make predictions.

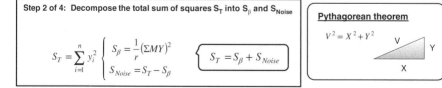

Step 2 of 4: Decompose the total sum of squares S_T into S$_\beta$ and S$_{Noise}$

$$S_T = \sum_{i=1}^{n} y_i^2 \begin{cases} S_\beta = \dfrac{1}{r}(\Sigma MY)^2 \\ S_{Noise} = S_T - S_\beta \end{cases}$$

$$S_T = S_\beta + S_{Noise}$$

Pythagorean theorem

$$V^2 = X^2 + Y^2$$

Figure 2.18   Step 2 of 4 for computations of S/N and β

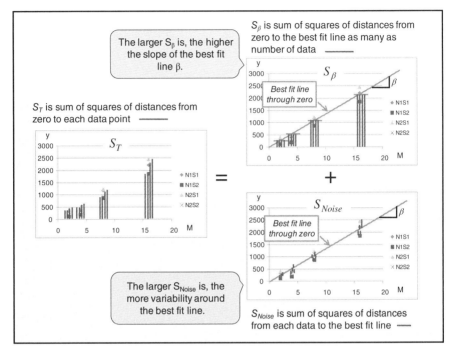

Figure 2.19 Graphical representation of decomposition of total sum of squares

- "Log" is multiplied by 10 because it becomes a number that is easier to handle.
- S/N has a value between $-\infty$ and $+\infty$. 2 dB is better than -5 dB; -5 dB is better than -12 dB.

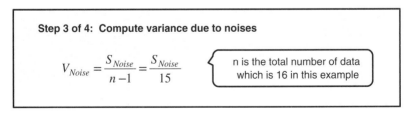

Figure 2.20 Step 3 of 4 for computations of S/N and β

Step 4 of 4: Compute S/N and β

$$\beta = \frac{1}{r}\Sigma MY \qquad S/N = 10\log\frac{\beta^2}{V_{Noise}} = 10\log\frac{\beta^2}{\sigma^2_{Noise}}$$

Figure 2.21   Step 4 of 4 for computations of S/N and β

2.2.6.2 Computation of S/N and β for L18 Data Sets

Table 2.10 shows the result of S/N and β computations for the entire L18. Figure 2.23 shows sample computation of S/N and β with No. 1 data set of L18.

The graphs in Figure 2.24 show the raw data plot for No. 1 and No. 14.

- S/N for No. 14 is $7.4 - (-3.9) = 11.7\,\text{dB}$ higher than No. 1.
- β for No. 14 is $(151.2 - 114.6)/114.6 \times 100 = 31.9\%$ higher than No. 1.
- No. 14 is definitely more robust than No. 1. We expect No. 14 feels better, has less quality problems and is more predictable and reliable.

2.2.6.3 Response Table for S/N and β

Table 2.11 is called a response table, which shows the average S/N and β for each factor's levels.

Computations to obtain values in a response table are quite simple. For instance, the average S/N for A1, denoted by $\overline{A}_1$, is the average of S/N whenever a test is run under A1. $\overline{A}_1$ is the average of S/N over Run No. 1 to No. 9. Similarly, $\overline{A}_2$, $\overline{B}_1$, $\overline{H}_1$, $\overline{H}_2$ and $\overline{H}_2$ are computed as follows:

Conceptually S/N is:

$$S/N = 10\log\frac{\beta^2}{\sigma^2_{Noise}} \quad (dB)$$

β = Efficiency of energy transformation

σ^2 = Variability of energy transformation

Figure 2.22   Concept of signal to noise ratio and $\tilde{\beta}$

Table 2.10 Computation of S/N and β for all 18 sets of data of L18

	A 1	B 2	C 3	D 4	E 5	F 6	G 7	H 8	r	MY	ST	Sβ	S Noise	V Noise	S/N	β
1	1	1	1	1	1	1	1	1	1360	155890	18355900	17868891	487008.8	32467.3	-3.9	114.6
2	1	1	2	2	2	2	2	2	1360	172570	22171000	21897357	273643.5	18242.9	-0.5	126.9
3	1	1	3	3	3	3	3	3	1360	181680	24350100	24270311	79789.4	5319.3	5.3	133.6
4	1	2	1	1	2	2	3	3	1360	167970	21036725	20745530	291194.9	19413.0	-1.0	123.5
5	1	2	2	2	3	3	1	1	1360	177240	23570250	23098542	471707.6	31447.2	-2.7	130.3
6	1	2	3	3	1	1	2	2	1360	175230	22666150	22577612	88537.6	5902.5	4.5	128.8
7	1	3	1	2	1	3	2	3	1360	166910	20587975	20484521	103454.3	6897.0	3.4	122.7
8	1	3	2	3	2	1	3	1	1360	126690	12061400	11801732	259667.6	17311.2	-3.0	93.2
9	1	3	3	1	3	2	1	2	1360	195300	28296050	28045654	250395.6	16693.0	0.9	143.6
10	2	1	1	3	3	2	2	1	1360	123680	11789225	11247605	541620.3	36108.0	-6.4	90.9
11	2	1	2	1	1	3	3	2	1360	169250	21889700	21062914	826786.4	55119.1	-5.5	124.4
12	2	1	3	2	2	1	1	3	1360	178750	23569725	23493796	75929.0	5061.9	5.3	131.4
13	2	2	1	2	3	1	3	2	1360	137394	14831746	13880229	951517.2	63434.5	-7.9	101.0
14	2	2	2	3	1	2	1	3	1360	205610	31147525	31084906	62619.0	4174.6	7.4	151.2
15	2	2	3	1	2	3	2	1	1360	143530	15894500	15147692	746808.2	49787.2	-6.5	105.5
16	2	3	1	3	2	3	1	2	1360	184770	25350275	25102907	247368.5	16491.2	0.5	135.9
17	2	3	2	1	3	1	2	3	1360	160700	19295525	18988596	306929.4	20462.0	-1.7	118.2
18	2	3	3	2	1	2	3	1	1360	151410	17730250	16856609	873641.1	58242.7	-6.7	111.3
														Average	-1.0	121.5

Numerical computation of S/N and β for No.1 of L18

		M1	M2	M3	M4	L
N1	S1	45	325	660	1635	32830
	S2	240	425	1020	1845	39860
N2	S1	40	340	810	1725	35520
	S2	290	575	1250	2175	47680
					SUM	155890

Linear tendencies

$$L_1 = 2(45) + 4(325) + 8(660) + 16(1635) = 32830$$
$$L_2 = 2(240) + 4(425) + 8(1020) + 16(1845) = 39860$$
$$L_3 = 2(40) + 4(340) + 8(810) + 16(1725) = 35520$$
$$L_4 = 2(290) + 4(575) + 8(1250) + 16(2175) = 47680$$

Sum of squares of M for all of 16 data

$$r = 4 \times (2^2 + 4^2 + 8^2 + 16^2)$$
$$= 1360$$

Sum of product of M and y

$$\Sigma MY = \sum_{i=1}^{4} L_i = 32830 + 39860 + 35520 + 47680 = 155890$$

The total sum of squares ST and its decomposition

$$S_T = \sum_{i=1}^{n} y_i^2 = 45^2 + 325^2 + \cdots + 2175^2 = 18355900$$

$$\begin{cases} S_\beta = \dfrac{1}{1360} 155890^2 = 17868891 \\ S_{Noise} = 18355900 - 17868891 = 487009 \end{cases}$$

Noise variance

$$V_{Noise} = \sigma^2_{Noise} = \frac{487009}{16-1} = 32467.3$$

β and S/N

$$\beta = \frac{1}{r} \Sigma MY = \frac{1}{1360} 155890 = 114.6$$

$$S/N = 10\log \frac{\beta^2}{\sigma^2_{Noise}} = 10\log \frac{113.6^2}{32467.3} = -3.93\, dB$$

r	MY	ST	Sβ	S Noise	V Noise	S/N	β
1360	155890	18355900	17868891	487008.8	32467.25	-3.93	114.63

Figure 2.23   Sample computation of S/N and β with No. 1 data set of L18

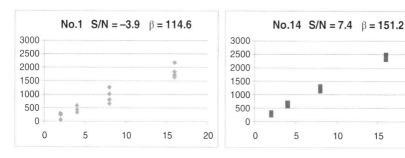

Figure 2.24   Raw data plot for high and low S/N

Table 2.11 Response tables for S/N and β

Response table (S/N)

	A	B	C	D	E	F	G	H
Level-1	0.3	-1.0	-2.6	-3.0	-0.1	-1.1	1.3	-4.9
Level-2	-2.4	-1.0	-1.0	-1.5	-0.9	-1.1	-1.2	-1.3
Level-3		-1.1	0.5	1.4	-2.1	-0.9	-3.2	3.1
Δ	2.7	0.1	3.0	4.3	1.9	0.2	4.4	8.0

Response table (β)

	A	B	C	D	E	F	G	H
Level-1	124	120	115	122	126	115	135	108
Level-2	119	123	124	121	119	125	116	127
Level-3		121	126	122	120	125	115	130
Δ	5.3	3.1	10.9	1.6	6.1	10.9	20.0	22.4

$$\overline{A}_1 = \frac{(-3.9)+(-0.5)+5.3+(-1.0)+(-2.7)+4.5+3.4+(-3.0)+0.9}{9} = 0.3\,dB$$

$$\overline{A}_2 = \frac{(-6.4)+(-5.5)+5.3+(-7.9)+7.4+(-6.5)+0.5+(-1.7)+(-6.7)}{9} = -2.4\,dB$$

$$\overline{B}_1 = \frac{(-3.9)+(-0.5)+5.3+(-6.4)+(-5.5)+5.3}{6} = -1.0\,dB$$

$$\vdots \quad \vdots \quad \vdots \quad \vdots \quad \vdots \quad \vdots \quad \vdots \quad \vdots \quad \vdots \quad \vdots \quad \vdots \quad \vdots$$

$$\overline{H}_1 = \frac{(-3.9)+(-2.7)+(-3.0)+(-6.4)+(-6.5)+(-6.7)}{6} = -4.9\,dB$$

$$\overline{H}_2 = \frac{(-0.5)+4.5+0.9+(-5.5)+(-7.9)+0.5}{6} = -1.3\,dB$$

$$\overline{H}_3 = \frac{5.3+(-1.0)+3.4+5.3+7.4+(-1.7)}{6} = 3.1\,dB.$$

Averages of β for each factor level are computed in a similar fashion. Δ in the table is the maximum difference among levels of each factor. The larger the value of Δ, the more influence the factor has. For instance, factor H has a huge 8.0 dB effect on S/N. Graphing factor effects make them more visible.

Response tables and graphs (Figure 2.25) show how control factors are affecting S/N and β. Let's observe what they are telling us. By the way, 1 dB gain in S/N is a significant effect. 0.1 dB gain is not significant at all. A gain of 6 dB is to reduce variation by 50%.

- Factor A: Pad Material affects S/N. A1: Type-1 is more robust than A2: Type-2 by 2.7 dB. This is a strong effect. A1 shows about 5% higher β as compared to A2. Since A1 is the baseline design, adapting A2 design improves robustness and efficiency.
- Factor B: Aspect Ratio does not affect S/N. B2 has slightly higher β than B1 and B3.

$$\vdots \quad \vdots \quad \vdots \quad \vdots$$
$$\vdots \quad \vdots \quad \vdots \quad \vdots$$

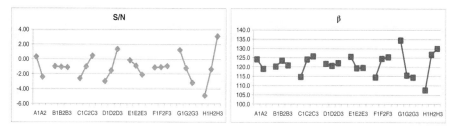

Figure 2.25 Response graph for S/N and β

- Factor H: Rotor has a huge effect on both S/N and β. H3 has a gain of 4.4 dB over the baseline H2. H1 is worse than the baseline by 3.6 dB.

By the way, don't celebrate yet. We cannot trust these effects unless we confirm them. We need to get a feel for how much we can trust this technical information. In order to get the confidence, we will make predictions for a couple of unknown combinations. Then data is collected from them. This way, we get to check how well the predictions meet the actual results. Those two unknown combinations are typically the optimum and the baseline.

2.2.6.4 Determination of Optimum Design

We now get to identify the optimum design. Two-step optimization is the basis of optimization.

Two-step Optimization

Step 1: Minimize the effect of noise/Reduce variability/Maximize S/N.
Step 2: Adjust the average response/Adjust β.

When it comes to "Energy," it is more difficult to reduce variability than adjusting the average. The strategy is to minimize variability first, then the average energy is adjusted to meet whatever the requirement is. Step 1 is to maximize S/N and Step 2 is to adjust β. In this case, during Step 1 it is desirable to have β as high as possible. Let's apply a two-step approach to select the optimum combination (Figure 2.26).

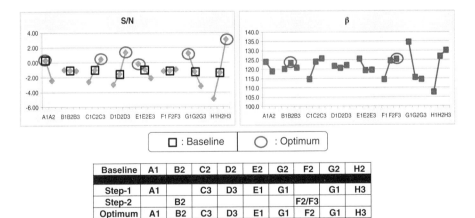

Baseline	A1	B2	C2	D2	E2	G2	F2	G2	H2
Step-1	A1		C3	D3	E1	G1		G1	H3
Step-2		B2					F2/F3		
Optimum	A1	B2	C3	D3	E1	G1	F2	G1	H3

Figure 2.26 Selection of optimum design

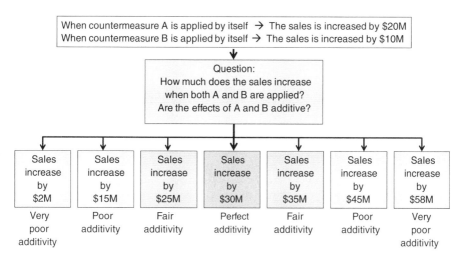

Figure 2.27   Concept of "additivity"

2.2.7 Step 7: Predict and Confirm

Before conducting "Confirmation," S/N and β are predicted for the optimum and the baseline. The prediction is made by assuming "Additivity" of factorial effects. Additivity simply means "Good + Good + Good = Excellent." Figure 2.27 illustrates the various levels of "Additivity" for control factors A and B attempting to increase company's total sales.

The following shows how predictions are made assuming "Additivity." Greek letter η is the notation for S/N. So we are to predict η and β, Eta and Beta, where η is "Robustness" and β is "Efficiency."

$$\begin{aligned}
\hat{\eta}_{Opt} &= \overline{T} + (\overline{A}_1 - \overline{T}) + (\overline{C}_3 - \overline{T}) + (\overline{D}_3 - \overline{T}) + (\overline{E}_1 - \overline{T}) + (\overline{G}_1 - \overline{T}) + (\overline{H}_3 - \overline{T}) \\
&= (-1.0) + (0.3 + 1.0) + (0.5 + 1.0) + (1.4 + 1.0) + (-0.1 + 1.0) + (1.3 + 1.0) + (3.1 + 1.0) \\
&= -1.0 + 1.3 + 1.5 + 2.4 + 0.9 + 2.3 + 4.1 \\
&= 11.5\,dB \\
\hat{\eta}_{Base} &= \overline{T} + (\overline{A}_1 - \overline{T}) + (\overline{C}_2 - \overline{T}) + (\overline{D}_2 - \overline{T}) + (\overline{E}_2 - \overline{T}) + (\overline{G}_2 - \overline{T}) + (\overline{H}_2 - \overline{T}) \\
&= -1.0 + 1.3 + 0.0 - 0.5 + 0.1 - 0.2 - 0.3 \\
&= -0.6\,dB
\end{aligned}$$

Here is the logic for making prediction assuming "Additivity." $\overline{T}$, the average of S/N from 18 rows, is used as the reference point. Then the dB gain from each factor from $\overline{T}$ is added together. $(\overline{A}1 - \overline{T})$ is the gain due to A1.

Notice that terms for factors B and F are missing in the prediction equations. The reason is to avoid over estimate. One easy way to avoid an over-estimate

Table 2.12 Summary of predictions

	Prediction	
	S/N	Beta
Base	-0.6	123
OPT	11.5	155
Gain	12.1	26%

is to discount weak effects in the prediction equations. The more that terms are discounted, the more conservative the prediction becomes. The less number of terms discounted, the more optimistic the prediction is.

Similarly, prediction for β is done by discounting factors B and F since B and F have weak effects on β. Predictions are summarized as shown in Table 2.12.

$$\hat{\beta}_{Opt} = \overline{T} + (\overline{A}_1 - \overline{T}) + (\overline{C}_3 - \overline{T}) + (\overline{E}_1 - \overline{T}) + (\overline{F}_2 - \overline{T}) + (\overline{G}_1 - \overline{T}) + (\overline{H}_3 - \overline{T}) = 123$$
$$\hat{\beta}_{Base} = \overline{T} + (\overline{A}_1 - \overline{T}) + (\overline{C}_2 - \overline{T}) + (\overline{D}_2 - \overline{T}) + (\overline{E}_2 - \overline{T}) + (\overline{G}_2 - \overline{T}) + (\overline{H}_2 - \overline{T}) = 155$$

By the way, the dB gain can be directly computed as follows.

$$\begin{aligned} dB\ Gain &= (\overline{A}_2 - \overline{A}_2) + (\overline{C}_3 - \overline{C}_2) + (\overline{D}_3 - \overline{D}_2) + (\overline{E}_1 - \overline{E}_2) + (\overline{G}_1 - \overline{G}_2) + (\overline{H}_3 - \overline{H}_2) \\ &= 0.00 + 1.5 + 2.9 + 0.8 + 2.5 + 4.4 \\ &= 12.1\ dB \end{aligned}$$

2.2.7.1 Confirmation

A prediction should be made and a confirmation should be run to check the validity of factorial effects (Figure 2.28). When it confirms the latter, we can trust the response tables as a trustworthy technical information.

2.2.8 Step 8: Lesson Learned and Action Plan

Rotating the Plan-Do-Check-Act cycle is critical in anything we do. This step closes the PDCA cycle (Figure 2.29).

- Reflect and list "lesson learned."
- Develop action plan.
- Communicate the result to all stakeholders.
- Document to the Corporate Memory.

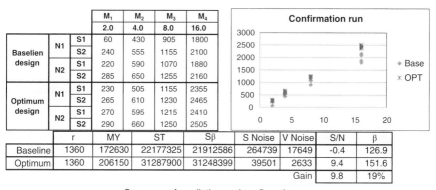

			M₁	M₂	M₃	M₄
			2.0	4.0	8.0	16.0
Baselien design	N1	S1	60	430	905	1800
		S2	240	555	1155	2100
	N2	S1	220	590	1070	1880
		S2	285	650	1255	2160
Optimum design	N1	S1	230	505	1155	2355
		S2	265	610	1230	2465
	N2	S1	270	595	1215	2410
		S2	290	660	1250	2505

	r	MY	ST	Sβ	S Noise	V Noise	S/N	β
Baseline	1360	172630	22177325	21912586	264739	17649	-0.4	126.9
Optimum	1360	206150	31287900	31248399	39501	2633	9.4	151.6
						Gain	9.8	19%

Summary of prediction and confirmation

	Prediction		Confirmation	
	S/N	β	S/N	β
Baseline	-0.5	122.9	-0.4	126.9
Optimum	11.5	154.7	9.4	151.6
Gain	12.0	26%	9.8	19%

Result of confirmation
- ❑ 9.8 dB Gain in Signal to Noise Ratio
- ❑ Efficiency increased 19%

> While the prediction was 12.0dB gain and 26% increase, the result of confirmation is not perfect, but sufficiently respectable.

Reduced weight and size
- ❑ Validation had verified "Squeal noise" reduced to 1/20 in its occurrence rate
- ❑ The optimized design is lighter, cheaper, quieter, more reliable, etc.

Figure 2.28   Results of confirmation and summary

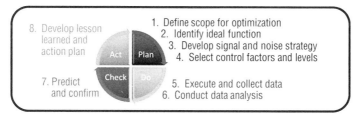

8. Develop lesson learned and action plan
7. Predict and confirm

1. Define scope for optimization
2. Identify ideal function
3. Develop signal and noise strategy
4. Select control factors and levels
5. Execute and collect data
6. Conduct data analysis

Act   Plan
Check   Do

Figure 2.29   PDCA cycle and 8 Steps of Optimization

2.3 Eight Steps for Robust Assessment

2.3.1 Step 1: Define Scope

As we discussed, Robust Assessment is to benchmark various designs with "Robustness" as criteria. Let us use the same brake subsystem example. The scope can be described as follows.

"Stopping distance of our current D-sedan has been inferior to key competitors. Customers are also complaining poor brake feel and confidence. In order to assure competitive performance for new D-sedan, we are to benchmark brake performance so we can set a strategic target. This study will benchmark the brake corner (Caliper, Pad and Rotor) performance to generate braking torque using dynamometer. Therefore, master cylinder, booster and tire are out of scope."

2.3.2 Step 2: Identify Ideal Function/Response and Step 3: Develop Signal and Noise Strategies

Step 2 is to decide what is measured and Step 3 is to identify under what customer conditions measurements are taken. These strategic steps are the same as Robust Optimization.

2.3.3 Step 4: Select Designs for Assessment

Because this is a benchmarking activity, there is no control factor. Some designs are strategically identified for benchmarking. For instance, they are: A1: Our Current Design, A2: Target Competitor's Design and A3: Best in Class. Step 1 to Step 4 are summarized in Figure 2.30.

2.3.4 Step 5: Execute and Collect Data

The ideal function is measured under noise strategy. Results are shown in Figure 2.31.

2.3.5 Step 6: Conduct Data Analysis

S/N and β are computed for A1, A2 and A3. The results are shown in Figure 2.32. Computations are the same as discussed in Step 6 of the Eight Steps for Robust Optimization (Section 2.2.6).

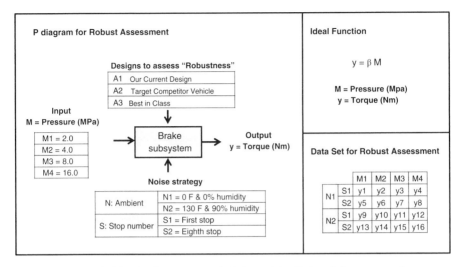

Figure 2.30   P-diagram and data set for Robust Assessment

2.3.6 Step 7: Make Judgments

Conclusions from the S/N and β analysis are very similar to what we concluded just by eyeballing raw data in the previous section. Here are some observations we can draw from this.

- The higher S/N, the more robust the performance is.
- The higher β, the more efficient the braking torque generation is. The higher the β, the shorter the stopping distance is.
- As compared to A1, A3 is more robust by 11.1 dB and A2 by 6.8 dB. A3's variability is about 1/4 of A1. A2's is less than 1/2 of A1.
- A3 produces 20% more braking torque than A1. A2 is 4% less than A1.
- Since A2 is our target, we need to improve S/N of our new design by 7 dB or higher while keeping the β at the current level, if not better.

2.3.7 Step 8: Lesson Learned and Action Plan

It is ideal to conduct Robust Optimization after Robust Assessment. When you find your design is better than any competitors, optimization can result in opportunities to reduce cost and weight. Strive for "Good to Great." In any case, continue rotating PDCA cycle. Moreover, optimization will result in less opportunities for symptoms of poor function. This does not mean you can omit the validation process at the end.

| | M1 = 2.0 | | | | M2 = 4.0 | | | | M3 = 8.0 | | | | M4 = 16.0 | | | |
| | N1 | | N2 | | N1 | | N2 | | N1 | | N2 | | N1 | | N2 | |
	S1	S2	S1	S2	S1	S2	S1	S2	S1	S2	S1	S2	S1	S2	S1	S2
A1: Our Current	110	320	165	395	180	430	320	500	860	1170	920	1215	1850	2210	1865	2445
A2: Target Competitor	200	265	230	285	450	595	510	620	870	1065	920	1005	1815	2020	1915	2115
A3: Best in Class	265	300	275	350	630	675	680	725	1185	1280	1290	1375	2395	2425	2410	2500

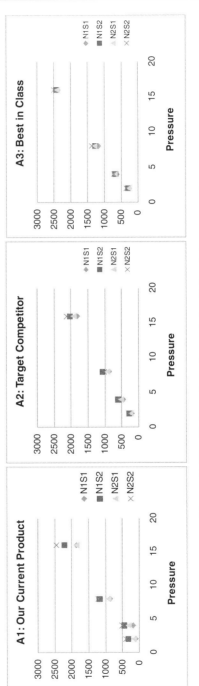

Figure 2.31   Data set and graph for Robust Assessment

	r	MY	ST	Sβ	S Noise	V Noise	S/N	β	S/N gain	β gain
A1: Our Current	1360	174940	23061425	22502944	558481.2	37232.08	-3.5	128.6	Base	Base
A2: Target Product	1360	167380	20708000	20600047	107952.6	7196.843	3.2	123.1	6.8	-4%
A3: Best In Class	1360	209940	32471000	32407944	63056.18	4203.745	7.5	154.4	11.1	20%

 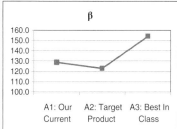

Figure 2.32 Computations of S/N and β for A1, A2 and A3 and plots

In this study, the action plan would be:

"Conduct Robust Optimization on our new design suing the same ideal function and noise strategy. The target for S/N is 4.0 dB or higher. The target for β is 130 or higher." After optimization the design must be validated.

2.4 As You Go through Case Studies in This Book

The Robust Optimization approach has been developed and improved for over last 30 years. Some older case studies may use equations that are different from those used in this chapter. For instance, the equations for S/N with $y = \beta M$ as the ideal function have evolved as shown below. The old ones are correct but complicated.

$$ S/N = 10 \log \frac{\frac{1}{r}(S_\beta - V_e)}{V_e} \quad \rightarrow \quad 10 \log \frac{\frac{1}{r}(S_\beta - V_e)}{V_{Noise}} \quad \rightarrow \quad 10 \log \frac{\beta^2}{V_{Noise}} $$

Because of the limitation of number of pages, many other types of S/N could not be covered. Those S/Ns include Standardized S/N and all non-dynamic S/N. Please refer to Taguchi's *Quality Engineering Handbook* by Genichi Taguchi, Subir Chowdhury, and Yuin Wu (Wiley, 2005) and the new ISO 16336: Robust Parameter Design.

3

Implementation of Robust Optimization

3.1 Introduction

In this chapter we introduce the implementation processes of robust optimization into the corporate environment and provide complete and detailed strategies behind the process. In it we will also show compatibility and possible enhancements of any other initiatives like Design For Six Sigma (DFSS) or Six Sigma that an organization may already have in place. Some organizations absolutely thrive on change and embrace the challenges and opportunities that come with change and diversity. Those organizations will have the foresight and vision to see that the robust optimization implementation strategy can be a guide for them and ultimately can help them to stay on the cutting edge and ahead of their competition.

3.2 Robust Optimization Implementation

The six critical components for a smooth implementation process for robust optimization are as follows:

1. Leadership commitment
2. Executive leader and the corporate team

Robust Optimization: World's Best Practices for Developing Winning Vehicles,
First Edition. Subir Chowdhury and Shin Taguchi.
© 2016 Subir Chowdhury, Shin Taguchi, and ASI Consulting Group, LLC.
Published 2016 by John Wiley & Sons, Ltd.

3. Effective communication
4. Education and training
5. Integration strategy
6. Bottom line performance.

It is extremely important that all of these elements be used. Additional favorable components may be present, of course, that complement corporate strengths and/or enhance and build on less developed areas of the organization. Companies that face major challenges and have the greatest needs for a working quality process will definitely have the most to gain from robust optimization.

3.2.1 Leadership Commitment

Robust optimization is one of the most powerful tools for making higher-quality products with little or no waste, and making them faster too. However, as with any new quality process, without the full support of senior leaders, the organization will not receive the full benefit of this program. Buy-in from top and middle management is essential. No matter how good or how efficient a program might be, it can be extremely difficult for personnel at the lower end of the organization if leadership is not committed to supporting it and making it work. It is essential that leaders fully understand all of the benefits of the process and see that the program is in no way a threat to their position or to their importance to the organization.

Today, any organization is dependent on the quality of its offerings. Leadership wants everything to be more efficient, faster, cheaper, and of higher quality. However, the only organizations that will achieve competitive status are those willing to invest in a more advanced and efficient methodology to achieve that quality.

When top management is not only supportive but encouraging, the rest of the organization will follow suit, and momentum will be gained for the process. A negative attitude by the powers-that-be can cause a breakdown of the program.

3.2.2 Executive Leader and the Corporate Team

Once all levels of management have bought into the idea of implementing robust optimization methodology, it is imperative that someone be chosen to be the driver, the leader of the initiative. This person should be a senior level manager directly reporting to Product Development Chief or CEO of the organization, highly efficient as a leader, motivated, and able to motivate others. He or she should be someone whom others would be honored to follow. He or she should be dependable and committed to the goals of the initiative, be firm in delegation, very detail-oriented, and strong on follow-up. Clearly, it is in the best interest of the organization that the corporate leader be highly respected and have knowledge of the work of all departments. He or she must be able to assert authority and have strong project-management skills. In short, you are looking for a "champion." Top leadership must be certain that they have chosen the right person for the job.

At the start of the process, an open meeting between upper management, middle management, and the chosen executive leader should take place. At this meeting, a complete understanding must be reached on all of the elements involved, and the tasks must be broken down to be more manageable and less overwhelming. If the methodology is not managed properly, the robust optimization initiative will not succeed.

The next most significant contributing factor to the success of the robust optimization implementation process is an empowered, multidisciplinary, cross-functional team. Since robust optimization implementation is in the product development area, it would be very beneficial to place as many engineers as possible on the team. The team must learn all of the techniques of robust optimization possible. The team should consist of representatives from every department and should be a mix of personalities and talents. It is also important to note, at this point, that there should be no cultural or organizational barriers that could cause the system to break down.

Each team member should have a clear understanding of the overall implementation process and of how his or her part contributes to the "big" picture. To this end, members of the team should divide and/or share all the tasks in the implementation according to responsibilities in the organization and personal abilities. Each team member should take full responsibility and be held accountable for his or her part in the process.

Defining the overall goals and the vision for the initiative will be the responsibility of a strategic planning team, and the executive leader should be actively involved with the strategic planning process as well as with the project. It is also the responsibility of each member of the cross-functional team to support, acknowledge, and communicate his or her understanding of the project and its objectives throughout the entire organization. Success of the process requires control, discipline, and some type of benchmarking against competitors as it progresses.

Managing change is difficult in any organization, and under most circumstances, employees have a natural resistance to and feel threatened by change. A project of the magnitude of robust optimization requires extremely effective management of resources and people to assure its success. Remember,

however, that should be considered part of the process. The project leader must report back to management on the process at each phase. If he or she sees that something is not moving as planned, he or she must work through members of the team or upper management to keep the process moving in the right direction. Often, when a breakdown occurs, it can be linked directly to unclear or ineffective communication.

Regarding successful deployment of robust optimization in original equipment manufacturers (OEMs) and suppliers, all of them being our clients in the global automotive industry, what we experienced was that in each case the executive leader, in most cases the Product Development Chief or EVP of engineering, was very passionate and involved on a daily basis with the implementation. The executive leader not only set up the vision, but was also supported with a budget for his or her direct reports. In almost all cases, the team supported each other and most importantly got themselves trained on the methodology itself.

3.2.3 Effective Communication

It is the responsibility of the cross-functional team to find ways to communicate with the entire organization. Every employee should receive the same messages, and the team must make sure that the meaning is clear to everyone. Robust optimization requires the organization to learn a new vocabulary, and it is the responsibility of the team to find interpretations that everyone can understand. The team members should consider what employees might ask about the process and then be prepared to answer those questions in language everyone will understand. Some of the questions that may come up are as follows:

- What is meant by *robustness*?
- What is meant by *optimization*?
- What is meant by *variability*?
- What is the robustness definition of failure?
- What is the operational meaning of prevention?

Of course, there are many other terms that must be defined. The task must be approached through well-thought-out, effective communication channels. Employees will react to what they perceive that a message says, so ambiguity must be avoided, and every base must be covered. The strategic planning team should put the necessary information into written form to help ensure that everyone is hearing and "getting" the same thing. Each cross-functional team member must see to it that his or her department members understand "what was said" as well as "what was meant." Every engineer and manager needs to be able to make decisions based on receiving the same information. Even the best of systems will fail if everyone is not pulling in the same direction, so clear, concise and timely information is crucial.

3.2.4 Education and Training

The importance of training and education can never be overstated, especially when new projects or initiatives are being introduced into an organization. A lack of understanding by even one employee can cause chaos and poor performance. The robust optimization process requires every employee to have an understanding of the entire process and then have a complete understanding of how his or her part in the process contributes to the bottom line. To that end, some training in business terminology and sharing of profit and loss information should be undertaken for every employee. Most times, robust optimization can be credited for cost avoidance such as less recalls or innovating a better design that may wow customers which in turn may get more market share. The purpose of doing robust optimization is promoting the fire prevention rather than fire fighting in the organization; which is a cultural mindset shift in most product development organization. Leaders must champion this idea of prevention and reward people for practicing robust optimization.

Training also can help to inspire and motivate employees to stretch their potential to achieve more. It is essential to an initiative such as robust optimization that everyone participate fully in all training. Engineers especially need to have a full understanding of the methodology and its potential rewards. But everyone, from top management to the lowest levels of the organization, must commit to focusing on the goals of the project and how they can affect the bottom line of either cost savings or cost avoidance. Knowledge adds strength to any organization, and success breeds an environment conducive to happier works.

Organizations the world over are discovering the value of looking at their organizations to find out what makes them work and what would make them work better. Through the development and use of an internal "expert" they are able to help their people work through changes, embrace diversity, stretch their potential, and optimize their productivity. These internal experts are corporate assets who communicate change throughout an organization and provide accurate information to everyone. Even when quality programs already exist, often the internal experts can work with the cross-functional team to communicate the right information in the right ways. These experts are also extremely efficient and helpful in spreading the goals and the purpose of robust optimization, as well as explaining the quality principles on which the methodology is built.

3.2.5 Integration Strategy

Trying to implement robust optimization methods and to create an environment where they can be seen as part of everyone's normal work activities is a challenge in itself. Many initiatives are in popular use in an organization, and some view robust optimization as just another program. However, such programs as quality function deployment (QFD), Pugh analysis, failure modes and effects analysis (FMEA), test planning, and reliability analysis are much more effective, take less time, and give higher performance with measurable results when combined with robust optimization.

Robust optimization is a unique approach to engineering thinking. The old way of thinking was "build it, test it, fix it." The robust optimization way is "optimize it, confirm it, verify it." The robust optimization implementation process has revolutionized the product development organization, and it is doing away with many of the more traditional tools and methods. Some companies, in part, have been able to change their perspectives on what constitutes failure and have been able to redefine reliability. When the emphasis is on variability, it can be associated with robust optimization.

When you make a product right in the first place, you don't have to make it again. Using the right materials and the right designs saves time, saves effort, saves waste, and increases profitability.

3.2.6 Bottom Line Performance

When one measures the results of new ways against the results of old ways, it is usually easy to tell which is better. When there are considerable differences, a closer look at the new way is certainly merited. Systems that implement robust optimization come out ahead on the bottom line. The real question is: Would you rather have your engineers spending their days putting out fires or would you rather have them prevent the fires by finding and preventing the causes? What if you offered incentives to your workers, the ones who actually made it all happen? What if you motivated them by rewarding them for differences in the bottom line? Your employees will feel very important to the organization when they see how they have helped to enhance the bottom line, but they will feel sheer ecstasy when they are able to share in the increased profits.

Part One
Vehicle Level Optimization

4

Optimization of Vehicle Offset Crashworthy Design Using a Simplified Analysis Model

Chrysler LLC, USA

4.1 Executive Summary

The task of developing crash worthiness for a new vehicle needs a systematic and calculated approach up to its production launch. Based on Voice of Customer (VoC) demands, the vehicle should be accurately sized with component optimization to insure consistent occurrences of structural and occupant responses during impact.

In the auto industry, virtual vehicle models, which are created using the computer aided engineering (CAE) tool, are widely utilized to evaluate and develop vehicle crash performance prior to tests and have proven to be an effective tool for improving vehicle designs. The vehicle stiffness demand ("demand" by kinetic energy, $\frac{1}{2} mv^2$) and supply ("supply" by energy absorption capability) planning needed in the early development stages

---

Robust Optimization: World's Best Practices for Developing Winning Vehicles,
First Edition. Subir Chowdhury and Shin Taguchi.
© 2016 Subir Chowdhury, Shin Taguchi, and ASI Consulting Group, LLC.
Published 2016 by John Wiley & Sons, Ltd.

requires engineers' prompt envisioning of all the impact design potentials. However, due to the significant development and analysis time requirements of the fully detailed CAE vehicle models, unique methodologies are required to support decisions on early design concepts.

This chapter presents a DFSS method application for a reliable and speedy vehicle crash stiffness demand/supply estimation in which a simplified spring-mass model is employed to expedite the crash performance prediction. This combination of the parameterized vehicle model and the DFSS optimization tool not only reduces the initial concept design period by 80% but also provides a much wider and in-depth insight of the causes and effects of the performance variations of vehicle crash designs en-route to the optimum one.

4.2   Introduction

Modern practice of vehicle crashworthiness development in the auto industry can be characterized by efforts toward speedy realization of a vehicle impact design that is fully optimized in overall crash functions including performance, impact robustness (repeatability), manufacturability, and light weight for fuel efficiency. This crash design process is usually implemented with a primary aim to satisfy the ever-changing government regulatory safety requirements. If a concrete direction for an optimum vehicle impact design can be established early in the program, it will pave the way to the final production design with minimal design changes. When the impact design process is in the beginning stage engineers have to envision all the possible cases of vehicle impact design concepts by using their cumulated expertise for the early design concepts. However, in the process of early optimum design selection among the candidate concepts, a large number of analysis model iterations or tests are required to simulate vehicle crash responses to design parameter changes. An analysis run of typical full system crash model takes 20–30 hours and a round of design optimization process handling 50(n) design control factors requires minimum $101(2n + 1)$ model iteration runs equivalent to 2000–3000 computer hours. Thus, early stage design optimization needs a unique analysis tool so that engineers can reduce the turnaround time for model analysis significantly.

This chapter has developed a new optimization method combining a simplified spring-mass model for analysis modeling and the DFSS technique for optimization processing. This type of study was conducted by Isuzu engineers [1] for the purpose of improving the vehicle full front impact (symmetrical loading case) performance. Our study, in contrast, approached optimizing the vehicle offset impact (asymmetrical loading case). The simplified spring mass models [2,3] were once a popular numerical analysis model in the early 1990s when the computer hardware was incapable of dealing with the modern complex vehicle analysis models consisting of almost one million finite elements. A drawback of spring mass models was compromised accuracy in results compared to full system model. Therefore, this study first improved accuracy and reliability of the spring mass model by employing the latest DYNA 3D nonlinear-spring and damper models [4], and achieved generating analysis results matching those of full vehicle models in accuracy and robustness enough for vehicle design optimization practice. This model has contributed to an 80% saving of the early design optimization process, enabling the difficult task of quantification of impact energy demand/supply for early concept design development.

4.3   Stepwise Implementation of DFSS Optimization for Vehicle Offset Impact

4.3.1   Step 1: Scope Defined for Optimization

- Study objective was defined to optimize vehicle offset impact design to achieve specific corporate performance targets at an early design stage.
- The team brainstormed performance control/noise factors through creating a Pugh chart.
- DFSS optimization was executed using a simplified spring-mass offset impact model for faster turnaround of the vehicle crash response.
- Optimized results were validated thoroughly based on the degree of their compliance with full system vehicle model impact responses.
- Identified optimal impact design concept was converted into optimization attributes of vehicle component design. (i.e., Material, Gage, Reinforcements, etc.).

4.3.2   Step 2: Identify/Select Design Alternatives

The team conducted selection criteria to evaluate design alternatives and implemented two rounds of Pugh matrix evaluations. Table 4.1 represents the Pugh matrix of round 1 design alternative selection for impact strategy.

From round 2 in which datum design was shifted from baseline to a better alternative identified in round 1, three design concepts were finally determined as best alternatives to pursue design optimization as shown in Table 4.2.

Table 4.1 Pugh matrix (round 1)

Attributes	Measurable Parameters	Current design		Body			Chassis		Body and chassis					Interior		Body and interior
		Baseline (Round1 DATUM)	New Strategy & within limits of tooling	Change (B1)	(B1)+ Change (B2)	Change (B3)	(C1) = New Chassis Concept 1	(C2) = New Chassis Concept 2	(B1)+(C2)+Enabler 1	(B1)+ (C2) + Enabler 2	(C2) with Enabler 3	(B1)+(B2)+(C1)	(B1)+(B2)+(C2)	Change (I1)	Change (I2)	(B1) + (I1)

Estimated degree of performance improvement

Assessment criteria

4.3.3 Step 3: Identify Ideal Function

For a dynamic response optimization with energy thinking [5, 6], two vehicle crash velocities (35 mph and 40 mph) were applied as input signals. The vehicle dash intrusions and the overall vehicle dynamic crush were chosen to be measured as output signals as shown in Figure 4.1. This output selection was based on the study objectives of maximizing engine box crush and minimizing dash intrusion.

Table 4.2 Pugh matrix (round 2)

Attributes		Measurable parameters	(B1) + (I1) (Round 2 DATUM)	(B1)+(I1)+ (C2) + Change (B3)	(I1)+(B1)+(C2)+Enabler 1	(I1) + Enabler 2	(I1) + (B1) + (C2)	(I1) + Modified Enabler 2	(I1) + (B1) + Modified Enabler 2

Estimated degree of performance improvement

Assessment criteria

Three concepts were chosen to optimize

P - Diagram

Spring model assessment

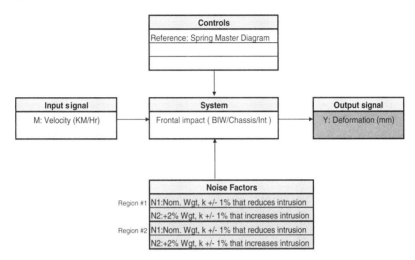

Figure 4.1   Program P-diagram

4.3.4 Step 4: Develop Signal and Noise Strategy

4.3.4.1 Input and Output Signal Strategy

Previously, the optimum design has been defined as one that generates soft crash pulse as well as sustains occupant compartment integrity during offset impact loaded events. To achieve this through DFSS, the two objective functions, one for maximizing displacement (VECD) of the vehicle engine compartment area (zone X1 in Figure 4.2) and the other for minimizing dash intrusions (VDI: crush of zone X2 in Figure 4.2), have been consolidated for computational simplicity by introduction of a new ratio factor $\boldsymbol{\beta_{1/2}}$ expressed by Equation (4.1) as follows:

$$\beta_{1/2} = \beta_1/\beta_2 \tag{4.1}$$

Now, a new consolidated objective function of maximizing the single output signal $\beta_{1/2}$ has been constituted, as subjected to crushable space

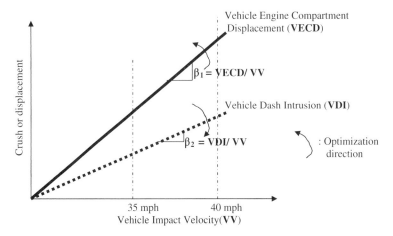

Figure 4.2   Ideal function schematic

constraints and as responded to the input signals of dual impact velocities. Figure 4.2 shows the schematic of the modified ideal function.

4.3.5 Step 5: Select Control/Noise Factors and Levels

4.3.5.1 Simplified Spring Mass Model Creation and Validation

It was impractical to perform this type of optimization process involving a large number of design parameters by using a full vehicle model which takes 20–30 hours per iteration run. Therefore, a spring-mass model having equivalent stiffness and mass to full system model has been developed by using Dyna3D [4] as shown in Figure 4.4. The simplified model consists of 32 nonlinear springs and 35 masses. Before it is used for

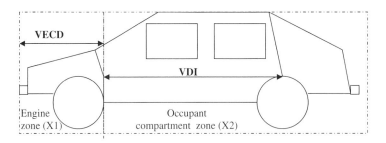

Figure 4.3   Vehicle crush zone classification

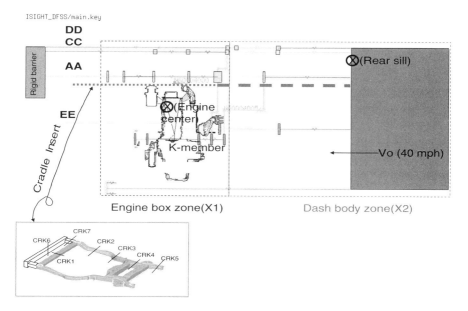

Figure 4.4 Spring mass model for vehicle offset impact event

optimization, the accuracy and reliability of its responses have been strictly validated by comparing with the full vehicle responses undergoing the same impact loading conditions. Figure 4.5 represents two of the response verification cases.

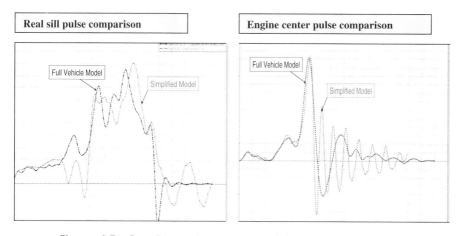

Figure 4.5 Baseline spring mass model response validation

4.3.5.2 Control Variable Selection

A total of 21 control factors, which were known to have influenced the vehicle offset impact performance most, were selected and given 3 control levels changeable from 0.9 to 1.1 as shown in Table 4.3. Then, through team brainstorming and engineering judgment, two levels of compounded noise factors, N1 and N2, were composed, of which N1 pertains to a noise condition which improves the output condition, and N2 pertains to a noise condition which degrades the output condition. The amplitude of 1% for the noise factor

Table 4.3 Spring mass model response validation

	Control factors (by factors to spring FD curves)				Noise factors (additional 1% up or down from selected level)	
ID	Spring ID's		Levels		N1	N2
A	DK2	0.9	1	1.1	1% Down	1% Up
B	DK3	0.9	1	1.1	1% Down	1% Up
C	DK4	0.9	1	1.1	1% Up	1% Down
D	DK5	0.9	1	1.1	1% Up	1% Down
E	CK4	0.9	1	1.1	1% Up	1% Down
F	CK5	0.9	1	1.1	1% Up	1% Down
G	AK3	0.9	1	1.1	1% Down	1% Up
H	AK4	0.9	1	1.1	1% Up	1% Down
I	AK5	0.9	1	1.1	1% Down	1% Up
J	AK6	0.9	1	1.1	1% Up	1% Down
K	AK7	0.9	1	1.1	1% Down	1% Up
L	EK2	0.9	1	1.1	1% Up	1% Down
M	EK3	0.9	1	1.1	1% Up	1% Down
N	KK1	0.9	1	1.1	1% Up	1% Down
O	KK3	0.9	1	1.1	1% Up	1% Down
P	KK4	0.9	1	1.1	1% Up	1% Down
Q	KK5	0.9	1	1.1	1% Up	1% Down
R	CRK1	0.9	1	1.1	1% Up	1% Down
S	CRK2	0.9	1	1.1	1% Up	1% Down
T	CRK3	0.9	1	1.1	1% Down	1% Up
U	KK4	0.9	1	1.1	1% Up	1% Down

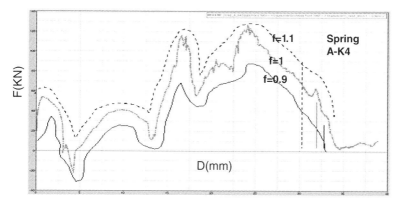

Figure 4.6 Control factor application for spring stiffness updating

variation was decided based on collected on-site data including such man-ufacturing variations as gage, material, and assembly.

4.3.5.3 Control Factor Level Application for Spring Stiffness Updates

Figure 4.6 illustrates an example of application of control factor levels (0.9, 1.0, or 1.1) to scale up or down the characteristic F-D (force-displacement) curve of the spring representing the factor ID.

4.3.6 Step 6: Execute and Conduct Data Analysis

An L54 array was employed to handle all the input data orthogonality requirements of the 21 control factors [5, 6]. A total of three rounds of optimization have been completed by replacing the nominal value of each control factor for current round of optimization with the previous round of optimization values. The amount of a mark-up or down for control factors was by 10% from round to round. Thus, this optimization has explored 35 billion design configurations ($=3^21 \times 3 + 3^19 \times 3$) that mean 21 control factors at 3 levels for 3 rounds of iterations for design concept I and 19 control factors at 3 levels for 3 rounds of iteration for design concept II respectively. Figure 4.7 shows the sensitivity of all the control factors with regard to output signal $\beta_{1/2}$, obtained from the first round of optimization, whereas Figure 4.8 illustrates the sensitivity data of control factors to the signal to noise ratio, ranging within $+/-0.5$ db variation interval. Results indicate that the noise factors N1 and N2 did not seem to cause severe variability allowing performance to deteriorate from the ideal function.

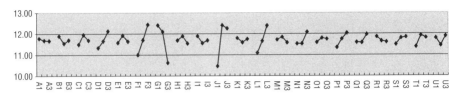

Figure 4.7   Control factor sensitivity to $\beta_{1/2}$ after first round optimization

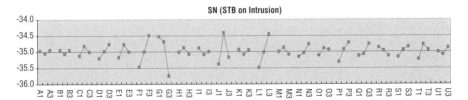

Figure 4.8   Control factor sensitivity to S/N after first round optimization

Figure 4.9 shows the optimum level choice of control factors, identified after Round 1 optimization. It was noted that influencing control factors were correlated between $\beta_{1/2}$ and S/N.

In Figure 4.10, dash intrusion, the major performance parameter, whose reduction was sought in this study, has reached the study target after application of the control factor levels of round 3 optimization results. The left side dash intrusion chart demonstrates the intrusion improvement status under the N1 noise conditions, while the right side indicates what occurred in N2 noise conditions.

4.3.7 Step 7: Validation of Optimized Model

Each of the control parameters, optimized up to the round 3 optimization, has been converted to the corresponding component stiffness of a full vehicle model. The impact responses of the full system model, enhanced with the

Overall optimum	A1	B1	C2	D3	E2	F3	G1	H2	I1	J2	K1	L3	M2	N3	O2	P3	Q3	R1	S3	T2	U1
Beta-1/Beta-2	A1	B1	C2	D3	E2	F3	G1,2	H2	I1	J2,3	K1,3	L3	M2	N3	O2	P3	Q3	R1	S3,2	T2,3	U3,1
SN (STB)	A1,3	B1,3	C2	D3	E2	F3	G1,2	H2	I1	J2	K1,3	L3	M2	N3	O2	P3	Q3	R1	S3	T2,3	U3,1

Figure 4.9   Table of round 1 optimization factor levels

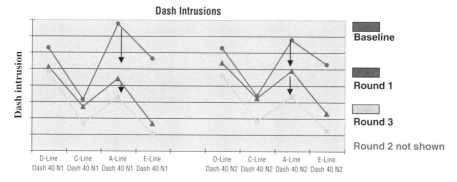

Figure 4.10   Dash intrusion improvement chart by three rounds of optimization

optimized structural strength, have been compared with the optimal responses of the round 3 simplified spring mass model as shown in Figure 4.11, Figure 4.12A and Figure 4.12B.

The first part of full vehicle validation was vehicle impact pulse comparison between the simplified spring mass model having the round 3 optimized spring F-D's and the full vehicle model with the converted round 3 optimized component stiffness as shown in Figure 4.12A. Notably, those two results are in good agreement.

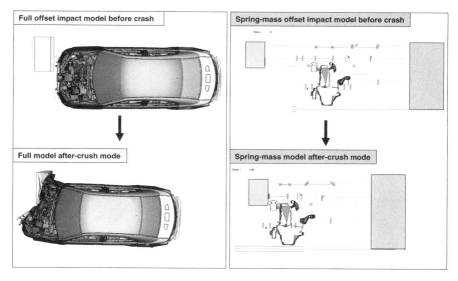

Figure 4.11   Full vehicle validation of round 3 optimized results

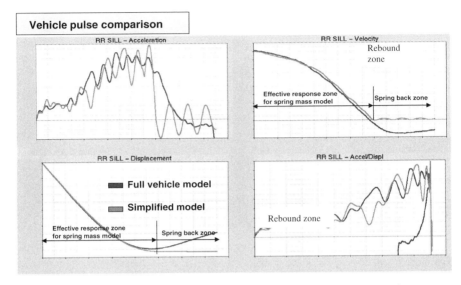

Figure 4.12A   Full vehicle validation of round 3 optimized impact performance

As the second part of full vehicle validation, vehicle dash intrusion comparison was made between the simplified spring mass model and the full vehicle model as shown in Figure 4.12B. Good match has been accomplished, thus the optimum stiffness enhancement package, identified by DFSS operation and simplified spring mass model, has been verified as enabling vehicle to perform to objectives.

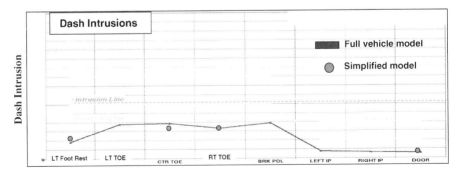

Figure 4.12B   Full vehicle validation of round 3 optimized dash intrusion

4.4 Conclusion

This study has successfully accomplished the objective of identifying a vehicle offset impact design concept whose crash performance in third-party rating and regulatory compliance tests has been verified through a full vehicle analysis model. In the process of this optimization, the coupling of the DFSS method and a simplified spring-mass model has been proven to be an effective and reliable optimization tool for faster development of impact design concepts at both component and vehicle levels. This optimization process enabled engineers to develop design concept in lesser time, avoiding late-program redesigns. The findings and benefits of this kind of DFSS optimization practice can be recapped as follows:

- Early stage vehicle crash design concept development turnaround time has been reduced by 80%. (Single analysis run time for simplified model of 10 minutes vs. full vehicle model run time of 30 hours.)
- Faster, more reliable optimization of vehicle crash performance and robustness is possible both in full vehicle level and component level designs.
- Tool for crash engineers to gain in-depth understanding of the interactive relations among the crash active and passive vehicle components during different impact loading conditions
- Vehicle dimension change scenario studies, which used to take a significant amount of time for model change of full vehicle analysis model, can be evaluated with minimal effort using simplified model.

4.4.1 Acknowledgments

The authors would like to thank Shin Taguchi for his assistance in conducting analysis data processing with the L54 orthogonal array.

4.5 References

1. Abe, M., Toyofuko, K., Fukunaga, S., and Watanabe, Y. (2005) Optimization of component characteristics for improving collision safety by simulation. *ASI's 21st Annual Symposium, Novi, MI, USA.*
2. Carrera, A., Mentzer, S., and Samaha, R. (1995) Lumped-parameter modeling of frontal offset impacts. SAE paper no. 95061.
3. Cheva, W., Yasuki, T., Gupta, V. *et al.* (1996) Vehicle development for frontal/offset crash using lumped parameter modeling. SAE paper no. 960437.

4. Livermore Software Technology Corp, 2004, *LS-DYNA 3D User's Manual*, Version 970.
5. Taguchi, G., Chowdhury, S., and Wu, Y. (2003) *Taguchi Quality Engineering Handbook*.
6. ASI Consulting Group, LLC (2007) *Design for Six Sigma Leadership Training Manual.*

This case study is contributed by Paul D. Duncanson, John A. DiGasbarro, Sae U. Park, and Kevin R. Thomson of Chrysler LLC, USA.

5

Optimization of the Component Characteristics for Improving Collision Safety by Simulation

Isuzu Advanced Engineering Center, Ltd, Japan

5.1 Executive Summary

The purpose of this study is the optimization of vehicle component characteristics to minimize the cabin deformation in collisions. A one-dimensional nonlinear spring and mass model was used to simulate a truck undergoing a frontal barrier crash test. There are many more variations between different truck models than between different passenger car models, including large differences in platform structure and payload capacity. The authors therefore carried out a parameter design process in which differences in component weights in different variations and changes in strength characteristics due to

Robust Optimization: World's Best Practices for Developing Winning Vehicles,
First Edition. Subir Chowdhury and Shin Taguchi.
© 2016 Subir Chowdhury, Shin Taguchi, and ASI Consulting Group, LLC.
Published 2016 by John Wiley & Sons, Ltd.

deterioration were considered. The parameter design process was iterated by checking closely spaced levels of the control factors to find the optimal direction. This method enabled the optimal conditions to be estimated while maintaining the reproducibility of the gain.

5.2 Introduction

A frontal barrier crash test, as shown in Figure 5.1, has been known as one of the test procedures of evaluating automotive crash safety performance. The objective of this research is the optimization of vehicle component characteristics for maximizing the occupant survival space of trucks in crash test. One-dimensional nonlinear spring and mass models have been used to simulate the frontal barrier crash test.

Each model of trucks has much larger number of variations than do passenger cars, and differences in cargo bed rigging and payload weights tend to be great. In our conventional approaches, we assumed conditions that could result in the smallest survival space of all cases, and then tried to minimize the cabin deformation under such conditions. The results, however, could be true to only one variation, and thus we were not quite sure whether they were effective for other variations as well. In addition, that practice was

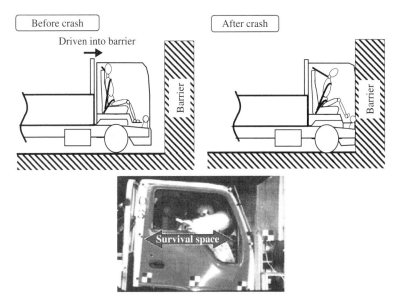

Figure 5.1 Frontal barrier crash test

quite time-consuming. In order to address these problems, the authors conducted a parameter design that takes account of differences in component weights resulting from different variations as well as changes in strength characteristics due to deterioration.

5.3 Simulation Models

For numerical simulation of major component behavior of vehicle in a barrier crash test shown in Figure 5.1, springs are assigned to parts subject to deform in crash, and masses are assigned to those not to deform as shown in Figure 5.2(a). For precise reproduction of crash events, a multiple-degrees-of-freedom model with increased number of masses and springs as shown in Figure 5.2(b) must be used. What the authors used in this research is a one-dimensional multiple-degrees-of-freedom model well capable of reproducing crash test results. As shown in Figure 5.3, nonlinear spring

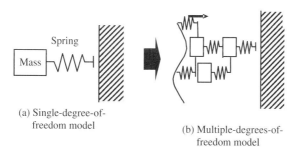

(a) Single-degree-of-freedom model

(b) Multiple-degrees-of-freedom model

Figure 5.2   Nonlinear spring-mass model

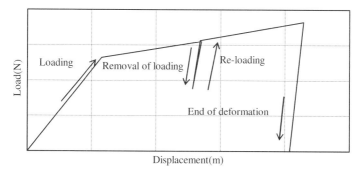

Figure 5.3   Example of spring deformation vs. load characteristics

characteristics capable of simulating plastic deformation are used. The objective of this analysis is to determine basic component structural characteristics in an early stage of designing where no detailed drawings are available. Using in-house analysis codes, we can easily change input/output relations. Calculation time has been significantly reduced to only several seconds for each case with this model. This analysis method is considered to be especially useful for research on variations because of the easiness in changing characteristic values.

5.4 Concept of Standardized S/N Ratios with Respect to Survival Space

As shown in Table 5.1, the input signals are crash velocity. Also shown in the table are the noise factors that represent variation in component masses and

Table 5.1 Level of factors

Types	Factors	Level 1	Level 2	Level 3
Signals	Crash velocity	Low V_1^*	Medium V_2^*	High V_3^*
Indicative	Payload weight	Light M_1^*	Medium M_2^*	Heavy M_3^*
Noise N* L_{36}	A′: Mass 1	Light	Heavy	
	B′: Mass 2	Light	Heavy	−
	C′: Mass 3	Light	Heavy	−
	D′: Mass 4	Light	Heavy	−
	E′: Mass 5	Light	Heavy	
	F′: Mass 6	Light	Heavy	−
	G′: Mass 7	Light	Heavy	−
	H′: Mass 8	Light	Heavy	−
	I′: e	−	−	−
	J′: Structural characteristics 1	−1%	+1%	−
	K′: Structural characteristics 2	−1%	+1%	−
	L′: Structural characteristics 3	−1%	0%	+1%
	M′: Structural characteristics 4	−1%	0%	+1%
	N′: Structural characteristics 5	−1%	0%	+1%

Table 5.1 (*Continued*)

Types	Factors	Level 1	Level 2	Level 3
Signals	Crash velocity	Low V_1^*	Medium V_2^*	High V_3^*
Indicative	Payload weight	Light M_1^*	Medium M_2^*	Heavy M_3^*
	O': Structural characteristics 6	−1%	0%	+1%
	P': Structural characteristics 7	−1%	0%	+1%
	Q': Structural characteristics 8	−1%	0%	+1%
	R': Structural characteristics 9	−1%	0%	+1%
	S': Structural characteristics 10	−1%	0%	+1%
	T': Structural characteristics 11	−1%	0%	+1%
	U': Structural characteristics 12	−1%	0%	+1%
	V': Structural characteristics 13	−1%	0%	+1%
	W': e	−	−	−
	A: Structural characteristics 1	−50%	0%	−
	B: Structural characteristics 2	−50%	0%	−
	C: e	−	−	−
	D: Structural characteristics 3	−50%	0%	+50%
	E: Structural characteristics 4	−50%	0%	+50%
	F: Structural characteristics 5	−50%	0%	+50%
	G: Structural characteristics 6	−50%	0%	+50%
Control L_{36}	H: Structural characteristics 7	−50%	0%	+50%
	I: Structural characteristics 8	−50%	0%	+50%
	J: Structural characteristics 9	−50%	0%	+50%
	K: Structural characteristics 10	−50%	0%	+50%
	L: Structural characteristics 11	−50%	0%	+50%
	M: Structural characteristics 12	−50%	0%	+50%
	N: Structural characteristics 13	−50%	0%	+50%
	O: Parts spacing 1	−50%	0%	+50%
	P: Parts spacing 2	−50%	0%	+50%

strength characteristic values, the control factors that represent strength characteristic, and the indicative factors that represent the payload weights.

The following are the two reasons for which signals are defined as crash velocity:

1. The kinetic energy of vehicle is dependent on the vehicle mass m and the crash velocity v ($E = mv^2/2$). When the kinetic energy before the crash becomes larger, the cabin deformation becomes proportionally large. Therefore, the kinetic energy could be handled as the signals. However, the author could not obtain a unique value when the signal factors are handled as the kinetic energy, since the vehicle component masses were defined as the noise factors.
2. The velocity is the vector. The signal is the scalar. Our simulation, however, uses one-dimensional model and thus the vector equals a simulated scalar, which in turn means no problem would result even if the velocity were used as the input.

The authors defined the payload masses as the indicative factor because the payload weight is just about the same as the vehicle weight, and is too large in variation to be defined as the noise factor. As Figure 5.4 indicates, the input signal represents crash velocity, and the output represents the magnitude of cabin deformation. The standardized S/N ratios have been determined on the basis of the standard condition N_{0i}^* representing the average mass of each component (A′–H′) and standard strength characteristics (T′–V′). L_{36} by L_{36} direct product test (simulation) has been conducted for each indicative factor. Table 5.2 lists the input signals, indicative factors, and output codes. The

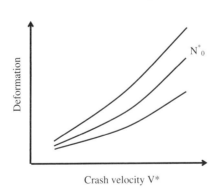

Deformation

N_0^*

Crash velocity V*

Figure 5.4   Input signals and output

Table 5.2 Input signals, indicative factors and output

		V_1^*	V_2^*	V_3^*
	N_{01}^*	$y_{0,11}$	$y_{0,12}$	$y_{0,13}$
M_1^*	N_1^*	$y_{1,1}$	$y_{1,2}$	$y_{1,3}$
	$\vdots$	$\vdots$	$\vdots$	$\vdots$
	N^*_{36}	$y_{36,1}$	$y_{36,2}$	$y_{36,3}$
	N_{02}^*	$y_{0,21}$	$y_{0,22}$	$y_{0,23}$
M_2^*	N_{37}^*	$y_{37,1}$	$y_{37,2}$	$y_{37,3}$
	$\vdots$	$\vdots$	$\vdots$	$\vdots$
	N^*_{72}	$y_{72,1}$	$y_{72,2}$	$y_{72,3}$
	N_{03}^*	$y_{0,31}$	$y_{0,32}$	$y_{0,33}$
M_3^*	N_{73}^*	$y_{73,1}$	$y_{73,2}$	$y_{73,3}$
	$\vdots$	$\vdots$	$\vdots$	$\vdots$
	N^*_{108}	$Y_{108,1}$	$y_{108,2}$	$y_{108,3}$

standardized S/N ratios have been calculated using these codes and the equations given below.

Effective divisors for each indicative factor:

$$r_1 = y_{0,11}^2 + y_{0,12}^2 + y_{0,13}^2 = 207597 \tag{5.1}$$

$$r_2 = y_{0,21}^2 + y_{0,22}^2 + y_{0,23}^2 = 449841 \tag{5.2}$$

$$r_3 = y_{0,31}^2 + y_{0,32}^2 + y_{0,33}^2 = 1426122. \tag{5.3}$$

Linear equation:

$$L_1 = y_{0,11} \times y_{1,1} + y_{0,12} \times y_{1,2} + y_{0,13} \times y_{1,3} = 165362 \tag{5.4}$$
$$\vdots$$

$$L_{108} = y_{0,31} \times y_{108,1} + y_{0,32} \times y_{108,2} + y_{0,33} \times y_{108,3} = 1408667. \tag{5.5}$$

Total variation:

$$S_T = y_{1,1}^2 + y_{1,2}^2 + \cdots + y_{108,2}^2 + y_{108,3}^2 = 76743955 (f = 324). \tag{5.6}$$

Variation of proportional term:

$$S_\beta = \frac{(L_1 + L_2 + \cdots + L_{107} + L_{108})^2}{36(r_1 + r_2 + r_3)} = 75949120 (f = 1). \tag{5.7}$$

Interaction between variation in proportional term and indicative factor:

$$S_{M^* \times \beta} = \frac{(L_1 + \cdots + L_{36})^2}{36r_1} + \frac{(L_{37} + \cdots + L_{72})^2}{36r_2} + \frac{(L_{73} + \cdots + L_{108})^2}{36r_3} - S_\beta$$

$$= 619(f = 2). \tag{5.8}$$

Interaction between variation in proportional term and noise:

$$S_{N^* \times \beta} = \frac{(L_1 + L_{37} + L_{73})^2 + \cdots + (L_{36} + L_{72} + L_{108})^2}{r_1 + r_2 + r_3} - S_\beta = 504769(f = 35).$$

$$\tag{5.9}$$

Error variation:

$$S_e = S_T - S_\beta - S_{N^* \times \beta} = 290066(f = 288). \tag{5.10}$$

Error variance:

$$V_e = \frac{S_e}{288} = 1007. \tag{5.11}$$

$$V_{N^*} = \frac{S_{N^* \times \beta} + S_e}{35 + 288} = 2461. \tag{5.12}$$

Standardized S/N ratio:

$$\eta = 10 \log \frac{(S_\beta - V_e)}{V_{N^*}} = 44.89 \, [\text{db}]. \tag{5.13}$$

5.5 Results and Consideration

Table 5.3 shows the ANOVA determined from the calculation with respect to the first line of the orthogonal array. In the applications of the Taguchi method, a wider range of level values has traditionally been recommended for both the control and noise factors, which is why we used ±50% as the level of the control factor in this study. This is the level acceptable in our design procedures. Figure 5.5 provides graphs of factorial effects. Control factors E, F, G, H, J, L, and O have ups and downs. As the results of calculation shown in Table 5.4 indicate, the gain is not properly reproduced. This is probably because the results are significantly affected by interactions.

Then, as suggested by Dr. Taguchi, the authors reduced the level range of Control factors to ±5%, which is one-tenth of the original range. This is

Table 5.3 ANOVA of the first line of the orthogonal array

Source	f	S	V
β	1	75949120	
M∗×β	2	619	
N∗×β	35	504769	
e	288	290066	1007
T	324	76743955	

Table 5.4 Summary of confirmatory calculation with wide level ranges [%]

	Estimated	Confirmed
Δ Comparison	40.43	40.44
○ Optimum	49.87	46.67
Gains	9.44	6.24

thought to be able to better use the advantage of simulation. For factors of which S/N ratios are subject to monotone increase, and which have large gains, we chose to find the trend of optimizing conditions by sliding the next levels in such direction to cause the S/N ratios to increase. Table 5.5 shows what happened when the levels have been changed six times, and Figure 5.6 shows how it caused the S/N ratio graphs of factorial effects to change. Shown in Table 5.6 are the projected and calculated values of both optimizing

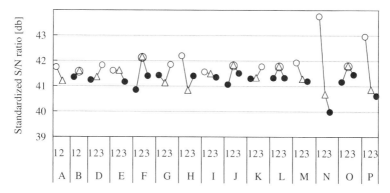

Figure 5.5 Standardized S/N ratio graphs of factorial effects with wide level ranges

Table 5.5 Processes of changing the levels of control factors six times

	Levels	1st	2nd	3rd	4th	5th	6th
A	1	−5%	←	←	←	←	←
	2	0%					
B	1	−5%	←	←	←	0%	←
	2	0%				+5%	
D	1	−5%	+5%	+15%	+25%	←	←
	2	0%	+10%	+20%	+30%		
	3	+5%	+15%	+25%	+35%		
E	1	−5%	−15%	−25%	−35%	−45%	←
	2	0%	−10%	−20%	−30%	−40%	
	3	+5%	−5%	−15%	−25%	−35%	
F	1	−5%	←	←	+5%	←	←
	2	0%			+10%		
	3	+5%			+15%		
G	1	−5%	←	←	←	+5%	+15%
	2	0%				+10%	+20%
	3	+5%				+15%	+25%
H	1	−5%	←	←	←	−15%	←
	2	0%				−10%	
	3	+5%				−5%	
I	1	−5%	←	←	+5%	←	←
	2	0%			+10%		
	3	+5%			+15%		
J	1	−5%	←	←	+5%	+15%	←
	2	0%			+10%	+20%	
	3	+5%			+15%	+25%	
K	1	−5%	+5%	←	←	+15%	+25%
	2	0%	+10%			+20%	+30%
	3	+5%	+15%			+25%	+35%
L	1	−5%	+5%	←	←	+15%	+25%
	2	0%	+10%			+20%	+30%
	3	+5%	+15%			+25%	+35%
M	1	−5%	−15%	−25%	←	←	−35%
	2	0%	−10%	−20%			−30%
	3	+5%	−5%	−15%			−25%
N	1	−5%	−15%	−25%	−35%	−45%	−55%
	2	0%	−10%	−20%	−30%	−40%	−50%
	3	+5%	−5%	−15%	−25%	−35%	−45%

Table 5.5 (*Continued*)

	Levels	1st	2nd	3rd	4th	5th	6th
O	1	−5%	←	←	←	←	←
	2	0%					
	3	+5%					
P	1	−5%	−15%	−25%	−35%	−45%	−55%
	2	0%	−10%	−20%	−30%	−40%	−50%
	3	+5%	−5%	−15%	−25%	−35%	−45%

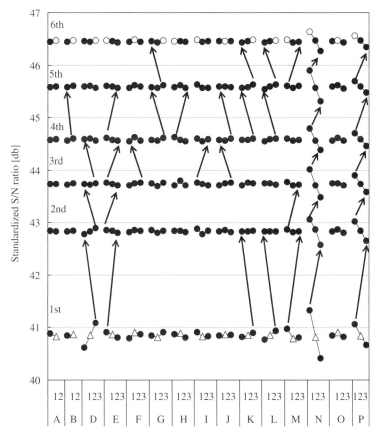

Figure 5.6 Standardized S/N ratios graphs of factorial effects for six level changes

Table 5.6 Summary of confirmatory
calculation for six level changes [db]

	Estimated	Confirmed
Δ Comparison	40.84	40.80
○ Optimum	46.94	46.73
Gains	6.10	5.93

and comparative conditions obtained after the sixth change. A good gain –
approximately 6 db – and good reproducibility have been obtained.

Here Table 5.7 shows how the optimum conditions compare wide control
factor level ranges with narrow ranges changed in six increments. Control
factors D, K, M, N and P, which show monotone increase or decrease with the
wide range, have the same direction of optimum values even in the latter with
the only exception of factor I which shows the reverse direction. And for
factors E, F, J, L and O, which have crestlike shapes, the 2nd level (0%) is
optimum in the case of wider level ranges, but not so in the narrow level
ranges, with the exception of factor O. Factor A shows widely different
optimum values for the wider level ranges and those calculated for each step.

Next, an attempt was made to determine whether the optimum conditions
obtained by changing the level ranges six times with narrow ranges would

Table 5.7 Comparison of optimum conditions [%]

	Wider level ranges	After 6th repetition
A	−50	0
B	0	5
D	50	35
E	0 (Crest)	−45
F	0 (Root)	5
G	50 (Root)	25
H	−50 (Root)	−15
I	−50	5
J	0 (Crest)	20
K	50	35
L	0 (Crest)	35
M	−50	−35
N	−50	−55
O	0 (Crest)	0
P	−50	−55

remain optimum in wide ranges of control as well. For this purpose, reproducibility was examined under the optimum conditions found after the 6th
repetitive calculation with wider level ranges of the Control factors. Table 5.8
shows the control factor level values determined using the optimum conditions obtained after changing the level ranges six times as the standard (0%).
The ranges of levels were set at ±50% except for factors that reach the design
limits. The signals and indicative and noise factors are the same as those
shown in Table 5.1 and thus are not given here. Figure 5.7 gives the S/N ratios
graphs of factorial effects, and Table 5.9 the results of the confirmatory
calculation. As is evident in these figure and table, the S/N ratio for the
optimum conditions is reproduced and the gain on the worst conditions is
also reproduced even with the wider level range. The authors have thus
confirmed that the method for finding optimum direction in steps using
narrow level ranges is effective to suppress the influences of interactions.

Table 5.8   Control factors with the optimum conditions as the standard (0%)

Types	Factors	Level 1	Level 2	Level 3
	A:Structural characteristics 1	−50%	0%	-
	B:Structural characteristics 2	−50%	0%	-
	C:e	-	-	-
	D:Structural characteristics 3	−50%	0%	15%
	E:Structural characteristics 4	−5%	0%	50%
	F:Structural characteristics 5	−50%	0%	40%
	G:Structural characteristics 6	−50%	0%	25%
Control L_{36}	H:Structural characteristics 7	−25%	0%	50%
	I:Structural characteristics 8	−50%	0%	45%
	J:Structural characteristics 9	−50%	0%	30%
	K:Structural characteristics 10	−50%	0%	15%
	L:Structural characteristics 11	−50%	0%	15%
	M:Structural characteristics 12	−15%	0%	50%
	N:Structural characteristics 13	0%	0%	50%
	O:Parts spacing 1	−50%	0%	50%
	P:Parts spacing 2	0%	0%	50%

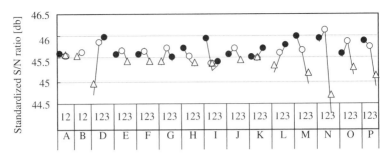

Figure 5.7 Standardized S/N ratios graphs of factorial effects with the optimum conditions used as the standard (0%)

Table 5.9 Summary of confirmatory calculation with the optimum conditions used as the standard (0%)

	Estimated	[db] Confirmed
△ Worst	41.75	42.25
○ Optimum	47.25	46.73
Gains	5.46	4.48

Finally, the sensitivities [1] determined from β_1, the primary coefficient of orthogonal expansion, are shown in Figure 5.8. The smaller the β_1, the better the results. The factor used here is the control factor L, which does not let the S/N ratio be affected by the optimum condition N_0^*, and thus permits proper

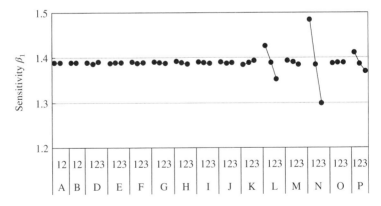

Figure 5.8 Graphs of factorial effects of sensitivity β_1

tuning. The tuning was made such that the magnitude of survival space deformation is no greater than the target value. The results are shown in Figure 5.9. In contrast with other conditions used for comparison, the optimum conditions after tuning are obviously smaller in variation, and the survival space deformation is held down below the target value. In conventional method of tuning, focus was given solely on sensitivity, thus the control

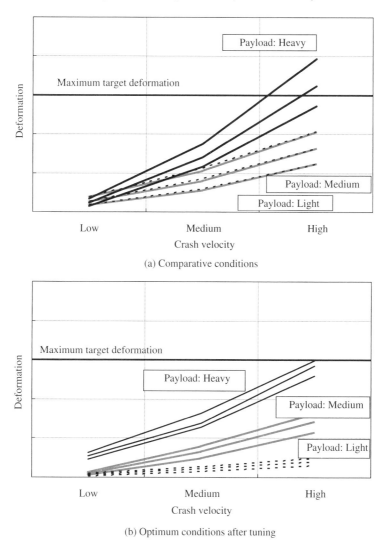

(a) Comparative conditions

(b) Optimum conditions after tuning

Figure 5.9   Final results of crash velocity vs. magnitude of cabin deformation

factor N was used. Using this factor to adjust the sensitivity affected the S/N ratio and the magnitude of deformation varied widely depending on varying conditions. Thus, in the conventional method, no indices were given to variations, which obviously made it very difficult to achieve the target values. Therefore, the two-step optimization is an effective method.

5.6 Conclusion

1. The optimum conditions have been properly estimated while maintaining proper gain reproducibility through the use of parameter design procedures in which optimization in multiple steps with narrow level value ranges was employed.
2. Application of quality engineering in early stages of development, where degrees of freedom are very small in simulation, turned out to be very effective.
3. A new design concept which is different from the traditional ones was found out.

5.6.1 Acknowledgment

The authors' gratitude is due to Mr. Hiroshi Yano, Ohken Associates, who provided us with generous guidance in carrying throughout our activities.

5.7 Reference

1. Taguchi, G. (2001) Objective functions and tuning, *Standardization and Quality Control*, **54** (8): 59–66.

This case study is contributed by Makoto Abe, Katsuya Toyofuku, Susumu Fukunaga, Yasuyuki Watanabe, of Isuzu Motors, Ltd, Japan.

Part Two
Subsystems Level Optimization by Original Equipment Manufacturers (OEMs)

6

Optimization of Small DC Motors Using Functionality for Evaluation

Nissan Motor Co., Ltd, Japan and Jidosha Denki Kogyo Co., Ltd, Japan

6.1 Executive Summary

In this study, the improvement of small DC motors used in automobiles was conducted using robustness as criteria. The robustness was evaluated by the relationship between the mechanical power and power consumption. It was considered ideal that the mechanical power required could be obtained from the minimum power consumption. The power consumption under the idle condition was also studied. After the improvement of robustness, audible noise level was significantly reduced. Also by utilizing part of the gain of robustness improvement for the downsizing of the motor, the cost and weight of the motor could be reduced.

Robust Optimization: World's Best Practices for Developing Winning Vehicles,
First Edition. Subir Chowdhury and Shin Taguchi.
© 2016 Subir Chowdhury, Shin Taguchi, and ASI Consulting Group, LLC.
Published 2016 by John Wiley & Sons, Ltd.

It is believed that the approach used in this study to evaluate the robustness of DC motor can be applied to the development of other types of motors towards the reduction of overall product development cost.

6.2 Introduction

Small DC motors are installed on many systems of a vehicle, including power windows and wipers, for example, and various kinds of quality improvement have been conducted such as audible noise reduction, vibration absorption, durability enhancement, etc. However such efforts have tended to focus on the single quality problem that was actualized then and to aim at meeting required spec or targets. Overall optimization for multiple quality problems has not been conducted efficiently.

In the near future, numerous electronic control units will be put into practical applications and load on in-vehicle battery tends to increase more and more. In order to respond to the various requirements on DC motors including low power consumption, new design optimization technology for DC motors has been needed at the R&D phase. Judging by these situations, robust engineering using robust assessment and optimization was applied to small DC motors to enhance global competitiveness in a shorter period of time.

6.3 Functionality for Evaluation in Case of DC Motors

A DC motor converts electrical power to mechanical power in order to drive objects. The mechanism of a DC motor can be expressed in the following expression:

$$2\,\pi\,nT = \beta'^{*}IE \text{ (unit:W).} \tag{6.1}$$

where

Electric Power [Current I (A)*Voltage E (V)]
Mechanical Power [2π*Revolutions n (rps)*Torque T (N-m)]

In this expression, the β' is equivalent to the energy efficiency and this expression indicates the technical means of a DC motor.

On the other hand, the requirement for a DC motor in actual use is to get mechanical power in order to drive windows or wipers. No matter how high the energy efficiency is, the motor that can't drive objects is useless. Given this

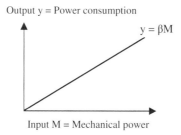

Output y = Power consumption

$y = \beta M$

Input M = Mechanical power

Figure 6.1   Ideal function of DC motor

factor, mechanical power required by a DC motor was defined as the signal factor and the ideal function of a DC motor was defined as gaining the required mechanical power with less power consumption (Figure 6.1).

The ideal function can be expressed in the following expression:

$$IE = \beta^{*} 2\pi\, nT\,(unit\!:\!W). \tag{6.2}$$

In the expression (6.2), β indicates the amount of power consumption to gain per unit mechanical power and can be defined as power consumption rate. Enhancing the robustness against noise factors and reducing the power consumption rate β in the expression (6.2) is equivalent to improving the essential functionality of DC motors.

6.4 Experiment Method and Measurement Data

In the experiments, revolutions and internal current were measured for 180 seconds at the sampling time of 0.1 second under the condition of impressed voltage corresponding to an in-vehicle battery with 3 load conditions corresponding to the objects to drive. The internal current under the unloaded condition was also measured and power consumption in idling was analyzed in order to confirm the possibility of DC motor optimization based on the idling data.

Figure 6.2 shows the temporal transition of power consumption under the each load condition for run No.1 of the L_{18} orthogonal array that is explained later. Since both the revolutions and the internal current were measured for 180 seconds at the sampling time of 0.1 second, 1800 data were recorded in the measuring machine for the each load condition. Judging from the temporal transition of data, it was assumed that 30 data, which are picked up from 3 fixed points of start-up, 90 seconds after start-up, 180 seconds after start-up could explain the overall trend.

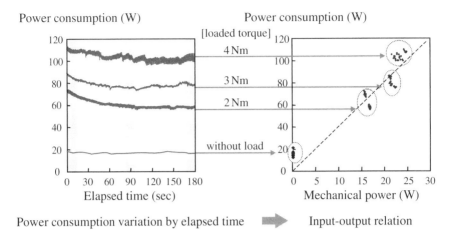

Power consumption variation by elapsed time ➡ Input-output relation

Figure 6.2   From measurement data to input–output relation

Therefore 10 data picking up at the sampling time of 1 second from the 3 fixed points ranged in 10 seconds were used for analysis. The number of data under each load condition was 30 and the total number of data under each run was reduced to120. The plotting result of the input–output relation from the 120 data is also shown in Figure 6.2.

The deviation from linearity in this relation corresponds to the variation of power consumption and nonlinear relation of the loaded torque and the internal current. These are equivalent to varriance due to noise.

6.5 Factors and Levels

The load torque was determined to be 3 levels of 2, 3 and 4 (N-m) that covered the operation range of the product. The revolutions were measured with the voltage of 12 (V) impressed, which is equivalent to an in-vehicle battery. The mechanical power was calculated by multiplying the load torque by the revolution and it was set as the level of signal factor. Since the revolutions were measured at the sampling time of 0.1 second, those varied with time due to the effect of internal heat generation even under the same load torque condition. Thus the levels of the signal factor varied at each measuring point. In this way, the number of levels on signal factor was equal to the number of data in each run. So the number of levels on signal factor is 90 excluding data of unloaded condition.

The noise factors are often selected from the environmental temperature or aging in the customers' usage conditions. However, in this experiment, the data

Table 6.1   Factors and levels

Factor \ Level	1	2	3
A : Type of fixing of A part	current*	rigid	---
B : Thickness of B part (mm)	0.15	0.17*	0.20
C : Shape of B part	current*	shape1	shape2
D : Width of D part (mm)	0.23	0.3	0.4*
E : Shape of E part	shape1	current*	shape2
F : Inner radius of F part (mm)	12.9*	13.5	14.0
G : Shape of G part (degree)	0*	9	18
H : Thickness of H part	1.0*	0.8	0.5
Signal M : Mechanical power (W)	Refer to Table 6.2		
Noise I : Elapsed Time	starting	1.5 min	3 min

The mark * indicates a current level

were intentionally measured under the continuous revolution that exceeded by far the actual usage conditions. Therefore the negative effects on energy efficiency due to the internal heat generation were induced as noise effects.

As explained above, internal heat generation due to excessive continuous revolutions was adopted as an effective noise strategy. Thus the test period based on the L_{18} array was reduced by approx. 50% in comparison with the typical assessment with the temperature conditions and the aging as noises in outer array. Eight control factors were selected. These factors were assigned on the L_{18} orthogonal array as shown in Table 6.3. Table 6.1 shows the factors and levels.

6.6 Data Analysis

The data was summarized in the table for each run of L_{18} orthogonal array as shown in Table 6.2. As previously explained, the motors were rotated for 3 minutes under 3 levels of load torque (T) while the impressed voltage was fixed to 12 V(E) and then revolutions (n) and internal current (I) were measured. Then, the mechanical power ($2\pi nT$) and power consumption (IE) were calculated at each measuring point. These values were defined as the levels of signal and responses, respectively.

Table 6.2 Experiment layout for each run

Operating condition		12V fixed									
	Voltage E										
	Torque T	T_0 (no load)			--	T_1 (2 Nm)			--	T_3 (4 Nm)		
	Time	0--3 min (0.1 sec S.T)			--	0--3 min (0.1 sec S.T)			--	0--3 min (0.1 sec S.T)		
		starting point	1.5 min passed	3 min passed	--	starting point	1.5 min passed	3 min passed	--	starting point	1.5 min passed	3 min passed
Measurement data	Data pick-up Rev :n	n_1--n_{10}	n_{11}--n_{20}	n_{21}--n_{30}	--	n_{31}--n_{40}	n_{41}--n_{50}	n_{51}--n_{60}	--	n_{91}--n_{100}	n_{101}--n_{110}	n_{111}--n_{120}
	Current :I	I_1---I_{10}	I_{11}---I_{20}	I_{21}---I_{30}	--	I_{31}---I_{40}	I_{41}---I_{50}	I_{51}---I_{60}	--	I_{91}---I_{100}	I_{101}---I_{110}	I_{111}---I_{120}
Signal : Mechanical P		0	0	0	--	2 n_{31}---	2 n_{41}---	--2 n_{60}	--	4 n_{91}---	4 n_{101}---	--4 n_{120}
Response : Electric P		12 I_1---	12 I_{11}---	12 I_{21}---	--	12 I_{31}---	12 I_{41}---	--12 I_{60}	--	12 I_{91}---	12 I_{101}---	---12 I_{120}

At first, the evaluation by one S/N ratio for both of the data under the loaded and unloaded conditions was sought. After the examinations, it has become clear that a few factors that reduce power consumption under the unloaded condition have a tendency to reduce the mechanical power under the loaded condition. It was assumed that a conflicting mechanism existed between power consumption reduction in idling and mechanical power enhancement under the loaded condition. Therefore the data of unloaded condition was analyzed separately from the data of loaded condition for additivity of factorial effects.

Since the signal and response are the attributes related to energy, each square root was used for analysis. This was done to enhance the additivity of the factorial effects obtained from the analysis based on the sum of squares. This is to say the square of square root of power becomes power itself.

Thus, for the 90 data of loaded condition, the square root of the signals: $2\pi n_{31}T_1, \ldots, 2\pi n_{120}T_3$, and the square root of the responses: $I_{31}E, \ldots, I_{120}E$ in Table2 were treated as $M_1, \ldots, M_{90}$ and $y_1, \ldots, y_{90}$, respectively.

The zero point proportional dynamic S/N ratio calculations were applied to the data.

$$S_T = y_1^2 + y_2^2 + \cdots + y_{90}^2 (f = 90) \tag{6.3}$$

$$r = M_1^2 + M_2^2 + \cdots + M_{90}^2 \tag{6.4}$$

$$L = M_1^* y_1 + M_2^* y_2 + \cdots + M_{90}^* y_{90} \tag{6.5}$$

$$S_\beta = \frac{L^2}{r}(f = 1) \tag{6.6}$$

$$S_e = S_T - S_\beta (f = 89) \tag{6.7}$$

$$V_e = \frac{S_e}{89} \tag{6.8}$$

$$\text{S/N ratio} \quad \eta = 10 * \log \frac{\frac{1}{r}(S_\beta - V_e)}{V_e} \tag{6.9}$$

$$\text{Sensitivity S} = 10 * \log \frac{1}{r}(S_\beta - V_e). \tag{6.10}$$

Since it is ideal that the power consumption is minimal in idling, the 30 data under the unloaded conditions were analyzed as the smaller-is-better characteristic as shown below.

$$\text{S/N ratio} \quad \eta = -10 * \log \frac{1}{30} \left[(I_1 E)^2 + (I_2 E)^2 + \cdots + (I_{30} E)^2 \right]. \tag{6.11}$$

6.7 Analysis Results

The values of the S/N ratio and the sensitivity for each run were calculated from the formula explained. These results are shown in Table 6.3 and the response graphs made from these results are shown in Figure 6.3 and Figure 6.4.

6.8 Selection of Optimal Design and Confirmation

From the response graph of the loaded condition (Figure 6.3), the levels with the higher S/N ratio showed the lower sensitivity except for the factors B and D. This result indicates that the smoother energy conversion causes the lower power consumption.

The trends of the factors B and H that greatly affect the power consumption under the unloaded condition are the same as the trends of the sensitivity of loaded data. These results are appropriate. On the other hand, the trends of the factors A, E and G slightly affect the power consumption under the

Table 6.3 Assignment and analysis result

Unit:db

No.	A	B	C	D	E	F	G	H	Loaded		Unloaded
									S/N ratio	Sensitivity	S/N ratio
1	1	1	1	1	1	1	1	1	11.20	6.00	−24.68
2	1	1	2	2	2	2	2	2	8.99	6.64	−23.84
3	1	1	3	3	3	3	3	3	14.61	5.99	−22.74
4	1	2	1	1	2	2	3	3	14.04	6.46	−23.04
5	1	2	2	2	3	3	1	1	9.33	6.65	−25.88
6	1	2	3	3	1	1	2	2	14.78	5.98	−24.85
7	1	3	1	2	1	3	2	3	11.95	6.21	−24.86
8	1	3	2	3	2	1	3	1	10.86	6.51	−26.36
9	1	3	3	1	3	2	1	2	9.72	6.81	−26.31
10	2	1	1	3	3	2	2	1	7.34	6.78	−24.77
11	2	1	2	1	1	3	3	2	12.22	6.47	−24.16
12	2	1	3	2	2	1	1	3	8.99	6.17	−22.09
13	2	2	1	2	3	1	3	2	11.90	6.21	−24.74
14	2	2	2	3	1	2	1	3	7.92	6.32	−22.97
15	2	2	3	1	2	3	2	1	12.54	6.66	−25.83
16	2	3	1	3	2	3	1	2	9.68	6.64	−24.81
17	2	3	2	1	3	1	2	3	14.92	6.20	−24.31
18	2	3	3	2	1	2	3	1	8.99	6.61	−26.71

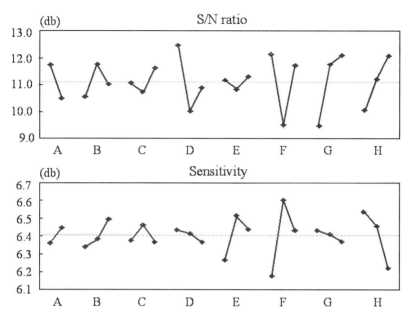

Figure 6.3   Response graph of S/N ratio and sensitivity analysis of data under loaded condition

unloaded condition are opposite to the trends of sensitivity under loaded condition. This means the power consumption reduction in idling causes the increase of power consumption rate under the loaded condition.

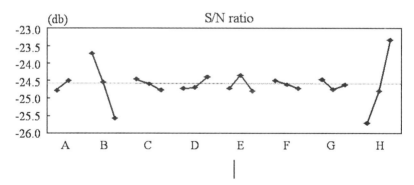

Figure 6.4   Response graph of S/N ratio: analysis of data under unloaded condition

Table 6.4 Predicted and confirmed results on S/N ration and sensitivity

Unit db

	Loaded condition				Unloaded condition	
	Predicted S/N ratio	Confirmed S/N ratio	Predicted sensitivity	Confirmed sensitivity	Predicted S/N ratio	Confirmed S/N ratio
Optimal	16.88	16.43	5.99	6.11	-23.29	-27.22
Current	10.06	11.73	6.31	6.93	-24.91	-29.73
Gain	6.82	4.70	-0.32	-0.82	1.62	2.51
Benchmark	----	16.77	----	6.62	----	----

Although the factors D and F show the significant effect on dynamic S/N ratio of the loaded condition, they show little effect on static S/N ratio of the unloaded condition. Considering these results, the dynamic S/N ratio of the loaded condition was prioritized for the selection of the optimal design that is $A_1 B_2 C_3 D_1 E_2 F_1 G_3 H_3$. In contrast, the current design is $A_1 B_2 C_1 D_3 E_2 F_1 G_1 H_1$.

The second level of factor E was selected for optimum based on the cost restrictions. Although the analysis results are not given here, the trend of the response graph was reproduced in confirmation tests of factor F, which showed V-shape factorial effect. Thus, the first level of factor F was selected for optimum.

Table 6.4 shows the predicted values of the S/N ratio and the sensitivity on the optimal design and the current design along with the results of the confirmation tests. As a result, good reproducibility could be confirmed for both the S/N ratio and sensitivity. In the confirmation test, the benchmark product that shows top-level performance in global market was evaluated.

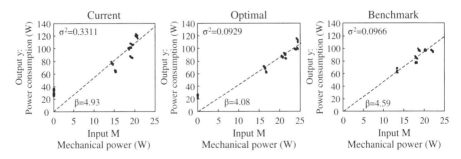

Figure 6.5 Comparison of input–output relation for each design

The results are included in Table 6.4. The S/N ratio on the optimal design was comparable to the benchmark product and the sensitivity was 0.5 db superior to the benchmark product.

Figure 6.5 shows the plotted input–output relations of the current design, optimal design and benchmark product. It is apparent that the optimal design has less variation of the power consumption than the current design. And the power consumption rate β on the optimal design have reached 4.08, which is approx. 17% lower than the current design of 4.93 and approx. 11% lower than the benchmark product of 4.59.

6.9 Benefits Gained

Through this optimization procedure, mechanical power required for the DC motor was obtained with less power consumption on the optimal design. This means that smooth energy conversion of the electric power and the mechanical power has been achieved. Therefore the side effects such as audible noise generated in the energy conversion can be expected to reduce. Since the energy efficiency of the optimal design is higher than that of the current design, the revolutions under the condition of the same voltage and rated load are also high. Therefore the audible noise was evaluated under the same revolutions as the condition of current design with an in-vehicle battery voltage and the rated load. The revolutions of the optimal design and the benchmark product were adjusted to the rated value by voltage regulation. As a result, the audible noise of the optimal design has been decreased approx. 8 dBA against the current design, achieving quietness equivalent to that of the benchmark product.

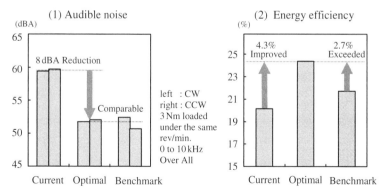

Figure 6.6 Performance comparison with benchmark product

Reduction of power consumption rate means an increase of mechanical power generated from the same electric power. Efficiency β' expressed in the expression of $2\pi nT = \beta'^* IE$ was enhanced by 4.3% compared to that of the current design and reached the excellent level exceeding that of the benchmark product by 2.7% (Figure 6.6).

Enhancement of the energy efficiency by 4.3% causes increase of the mechanical power obtained from an in-vehicle battery. The mechanical power of the optimal design has reached 120% in comparison with the current design under the same load condition. This 20% gain of mechanical power can be interchanged with the cost reduction through motor downsizing. While the mechanical power equivalent to that of current design is maintained, motor parts such as magnet, yoke, etc. can be downsized on the optimal design. The effect of downsizing is almost equivalent to the cost reduction of 250 000 dollars a year mainly through the reduction of material cost.

6.10 Consideration of Analysis for Audible Noise

The study so far verified improvement of robustness of DC motor functionality brought the enhancement of energy efficiency, audible noise reduction and cost reduction. On the other hand, it is said that the optimal design based on only the evaluation of quality characteristics like an audible noise will bring many problems. In order to confirm what kind of problems arise, analysis of the audible noise for each L_{18} array and comparison of the analysis result by robust optimization was conducted.

In case of a brush motor, audible noise is compounded with magnetic noise and brush friction noise and it is possible to divide them based on frequency characteristic and analyze them separately. But in this study, complex sonic pressure (Pa) of the overall value from 0 kHz to 10 kHz was analyzed as a smaller-is-better characteristic taking up the rotating directions as the noise factor. The response graphs obtained from the sonic pressure analysis and the analysis on robustness are shown in Figure 6.7.

Comparing these response graphs of functionality and sonic pressure, factors A, B, C, D and E showed the direct opposite trend. These results indicate the selection of the best levels for these factors based on the analysis of sonic pressure is equivalent to the selection of the worst levels for robustness of motor function.

In order to confirm the fact, the predicted value of S/N Ratio and sensitivity are shown in Table 6.5 concerning the optimal design for robustness ($A_1 B_2 C_3 D_1 E_2 F_1 G_3 H_3$) and the noise-oriented design determined from the sonic

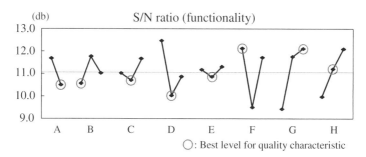

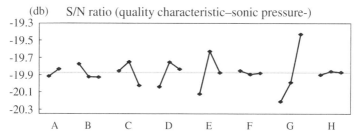

Figure 6.7 Functionality vs. quality characteristic for evaluation

pressure analysis (A_2 B_1 C_2 D_2 E_2 F_1 G_3 H_2) along with the current design (A_1 B_2 C_1 D_3 E_2 F_1 G_1 H_1).

It was predicted that the noise-oriented design achieved the lower audible noise level compared with the optimal design. On the other hand, the noise-oriented design was predicted to be worse in energy efficiency compared not only with the optimal design but also current design. One of the technical reasons why this phenomenon occurs is that if the brush contact pressure is

Table 6.5 Comparison between the optimal design of functionality and the noise-oriented design

	Analysis of audible noise	Robustness			Predicted efficiency (%)
	Predicted S/N ratio (db)	Predicted S/N ratio (db)	Predicted sensitivity (db)	
Optimal design	−19.55	16.88	5.99	25.25
Noise-oriented design	−18.75	10.29	6.35	22.95
Current design	−19.99	10.06	6.31	23.15

eased, friction noise will be reduced; however, the loss in energy flow will increase and then energy efficiency will get worse.

Thus more input energy will be needed to obtain the same required mechanical power under the noise-oriented design as that under the current design. So under the same output condition of mechanical power, the audible noise level under the noise-oriented design will get worse than that under the current design. This result verifies the famous saying of Dr. Taguchi, "To get quality, don't measure quality."

In other words, it is extremely inefficient to solve the various quality problems occurring in the mechanical system one by one. To achieve dramatic improvement of the efficiency in R&D phase, the concept of evaluating and improving the functionality that is unique to each system is much more effective for solving a variety of quality problems at once including the unpredictable problems.

6.11 Conclusion

6.11.1 The Importance of Functionality for Evaluation

In this study, the relation between mechanical power and power consumption was evaluated to improve the robustness of DC motor function. As a result, a greater improvement was achieved than initially expected. Through enhancing robustness, significant reduction of side effects like an audible noise and cost reduction due to the downsizing was achieved.

This is the great improvement that has not been conventionally achieved from the evaluation of quality characteristics using the one-factor-at-a-time experiment.

6.11.2 Evaluation under the Unloaded (Idling) Condition

The possibility of improving the DC motor based on the idling data was also studied after measuring power consumption under the unloaded condition. As a result, some factors showed the trends that reduction of power consumption under the unloaded condition caused mechanical power reduction under the loaded condition. It could be judged that the optimization exclusively under the unloaded condition was difficult. This results mean, once again, that it is important to set the load torque as the signal factor, which is corresponding to the usage condition of customers when DC motor optimization is conducted effectively.

6.11.3 Evaluation of Audible Noise (Quality Characteristic)

It has been verified that assessing robustness of energy transformation is important for efficient improvement of the total system. In this study, focusing on audible noise, which is one of the quality characteristics, the problems and inefficiency caused by evaluation of quality characteristic were studied. As a result, the noise-oriented design determined from the sonic pressure analysis achieved the superior audible noise level to the optimal design. But the mechanical power obtained from an in-vehicle battery was significantly reduced. Therefore the noise-oriented design needs the higher input energy in order to obtain the required mechanical power. So then, under the same output condition of mechanical power, it was expected that the audible noise of the noise-oriented design should become worse than that of the current design. On account of this, the importance of evaluating the functionality that is unique to each system was reconfirmed.

After this study, the energy efficiency of the optimal design could be superior to that of the benchmark product that has been a long-standing target. And the audible noise level of the optimal design was less than or comparable to that of the benchmark product. As a result, the balance of quality and cost was greatly improved and globally top-level performance has been achieved.

In the immediate future, we plan to apply this robust optimization approach to the development of other types of motors in order to shorten the lead time of R&D phase and reduce the total development costs.

6.11.4 Acknowledgment

Finally, we are deeply grateful to Dr. Genichi Taguchi for his considerate guidance on this study.

This case study is contributed by Kanya Nara and Kenji Ishitsubo of Nissan Motor Co., Ltd, Japan and Mitsushiro Shimura and Hideki Rikanji of Jidosha Denki Kogyo Co., Ltd, Japan.

7

Optimal Design for a Double-Lift Window Regulator System Used in Automobiles

Nissan Motor Co., Ltd, Japan and Ohi Seisakusho Co., Ltd, Japan

7.1 Executive Summary

There has been an increasing trend of using double-lift window regulator systems on luxury automobiles with the aim of improving regulator perform-ance. In this study, the parameter design was used to optimize the window regulator system and improve the robustness of the system in the market. Besides the analysis based on the conventional dynamic S/N ratio, the analysis based on the standardized S/N ratio was used in order to compare two analyses. It was confirmed, as shown in this study, that there is little difference in the results obtained with two analyses when the linear term is significant.

Robust Optimization: World's Best Practices for Developing Winning Vehicles,
First Edition. Subir Chowdhury and Shin Taguchi.
© 2016 Subir Chowdhury, Shin Taguchi, and ASI Consulting Group, LLC.
Published 2016 by John Wiley & Sons, Ltd.

7.2 Introduction

Window regulator systems that use the power of motor to lower and raise the window glass are widely used in automobiles today. Because these systems are used so frequently in usual driving and operate very close to customers, it is required that the system does not give rise to any quality problem. Such problems include rattles, changes in operating speed, squeaks, stopping of the window at midway and so on, and these requirements are especially severe in the market of luxury automobiles. Automotive manufacturers are competing fiercely to develop new window regulator systems, in addition to improve their existing systems.

Previously, single-lift window regulator systems, which support the window glass at one point, were widely used. However, double-lift window regulator systems that support the window glass at two points have been increasingly adopted in luxury automobiles in recent years to improve operating performance. In this study, we applied the parameter design to optimize the double-lift window regulator system with the aim of reducing troubles in the market to zero.

In addition, the analysis was also conducted based on the standardized S/N ratio. The advantages of using the standardized S/N ratio were examined by making a comparison between the results based on the standardized S/N ratio and the conventional S/N ratio.

7.3 Schematic Figure of Double-Lift Window Regulator System

The double-lift window regulator system examined in this study is shown schematically in Figure 7.1. The window glass is supported at two points, that is, at the front and the rear, by the cable that lowers and raises the window glass via clamps, which move on the front and the rear rail. The cable is driven by a motor. The glass run is also located between the window frame and the edge of the glass to ensure water tightness and air tightness.

7.4 Ideal Function

The purpose of a window regulator system is to move the window glass from a certain position to another desired position. Accordingly, for the

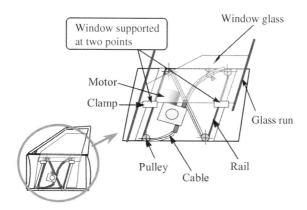

Figure 7.1   Schematic figure of double-lift window regulator system

ideal function of the system, we considered that its operating time should be proportional to the distance, which the window glass moves on the rails, as shown in Figure 7.2. Additionally, it is desired that this relationship is not influenced totally by noise factors and also the relationship maintains a high degree of linearity. When the window regulator system is considered in terms of its application to automobiles, the lowering and the raising speed of the window glass has a target value. However, with respect to the mechanical operation of the regulator system itself, the relationship between speed and distance of movement should have a small value of the slope β. In other words, excellent efficiency is expected for an ideal window regulator system.

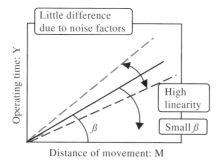

Figure 7.2   Concept of ideal function

Actual noise factors
causing variations

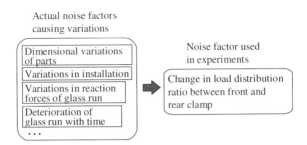

Noise factor used
in experiments

Dimensional variations
of parts

Variations in installation

Variations in reaction
forces of glass run

Deterioration of
glass run with time
...

Change in load distribution
ratio between front and
rear clamp

Figure 7.3   Noise factor used in experiments

7.5 Noise Factors

Various noise factors affecting the operating performance of the window
regulator system are listed in Figure 7.3. These factors include dimensional
variations of constituent parts, variations in the installation of the front and
the rear rail, variations of the reactive force of the glass run and its change with
time, and variations in sliding resistance due to environmental conditions
such as temperature and humidity. All of these factors have the effect of
varying the magnitude of the load (resistance) placed on the front and the rear
clamp and the ratio of the load distribution between them.

Considering experimental efficiency, it was decided not to incorporate each
of these factors causing variations directly into the experiments. Instead, an
experimental jig was designed, as shown in Figure 7.4 in which a beam was
used to connect the clamps without putting the window glass in clamps. 12 kg

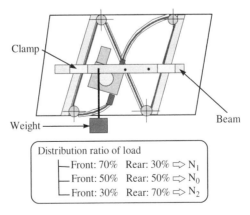

Clamp

Weight

Beam

Distribution ratio of load
— Front: 70%   Rear: 30% $\Rightarrow N_1$
— Front: 50%   Rear: 50% $\Rightarrow N_0$
— Front: 30%   Rear: 70% $\Rightarrow N_2$

Figure 7.4   Way of varying load distribution

weight, corresponding to the weight of the window glass combined with the resistance of the glass run, was used as the substitute for many noises by varying the position where the weight was attached to the beam.

The weight distribution ratio under the standard design condition N_0 was 50% at the front and 50% at the rear. It is known from previous data that the range of variation produced by the above-mentioned noises is ±20%. Depending on the mechanism of the regulator operation, it is also known that resistance increases when the load distribution is offset toward the front clamp and decreases when it is offset toward the rear clamp. Therefore, two noise conditions were defined for the load distribution, that is, the N_1 with the front/rear split of 70/30% and the N_2 condition with the split of 30/70%. Aiming at that the difference of the performance between the N_1 and the N_2 condition would be small and the mean value of the slope β of the N_1 and the N_2 conditions would be small, the optimal design specification was pursued.

The standard battery voltage used to drive the motor of window regulator systems is 13.5 V, but voltages in automobiles of the market can vary in the range of 11–15 V depending on the using conditions of other electrical equipments at the time. Accordingly, battery voltage is also a factor that causes variations. However, a constant voltage of 9.4 V was used in preliminary experiments, even in this low voltage the window regulator system could be operated under the worst possible condition. The battery voltage of the experiment was kept constant in order to evaluate more clearly the difference between a good design of a window regulator system and a poor design.

It was also revealed in preliminary tests that virtually no differences between a good design and a poor regulator design could be seen in the downward movement of the window glass, owing to the large effect of the gravity by the weight of the window glass itself, compared with upward movement. Accordingly, even though the window glass moves upward and downward actually, evaluations focused only on the upward movement of the window glass.

The distance the window glass moved was divided equally into three levels that were used as the signal factor.

7.6 Control Factors

Seven factors were chosen as control factors and assigned to an L_{18} orthogonal array, which were thought to have the effect for the operating performance of the window regulator system. The third column of the array was left empty.

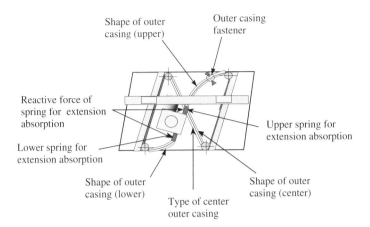

Figure 7.5   Control factors

Two levels were set for the control factor E (the fastened condition of the outer casing) and the control factor F (the type of the center casing) and the third level was treated as a dummy. Control factors are shown in Figure 7.5, and factors and levels used in the analysis are given in Table 7.1.

Table 7.1   Factors and levels

		Factors	Level 1	Level 2	Level 3
Control factor	A	With/without spring for extension absorption	With*	Without upper	
	B	Shape of outer casing (center)	Large radius	Middle radius*	Small radius
	C	Shape of outer casing (upper)	Large radius*	Middle radius	Small radius
	D	Shape of outer casing (lower)	Large radius*	Middle radius	Small radius
	E	Outer casing fastener	Without*	With	(Without)*
	F	Type of center outer casing	Type A*	Type B	(Type A)*
	G	Reactive force of spring for extension absorption	Small	Middle	Large*
Signal factor	M	Distance of movement	130 mm	260 mm	390 mm
Noise factor	N	Load distribution	Front: 70% Rear: 30%	Front: 30% Rear: 70%	—

*Initial Condition

7.7 Conventional Data Analysis and Results

The data obtained in experiment No. 1 are shown in Table 7.2. Using these data, the S/N ratio and the sensitivity were found based on the following procedures.

Total sum of squares:

$$S_T = 7.17^2 + 11.52^2 + \cdots + 12.17^2 = 682.7856 \quad (f = 6).$$

Effective divisor:

$$r = 130^2 + 260^2 + 390^2 = 236600.$$

Linear form:

$$L_1 = 130 \times 7.17 + 260 \times 11.52 + 390 \times 15.66 = 10034.70$$
$$L_2 = 130 \times 5.23 + 260 \times 8.83 + 390 \times 12.17 = 7722.00.$$

Variation of proportional term:

$$S_\beta = \frac{(L_1 + L_2)^2}{2r} = 666.3153 \quad (f = 1).$$

Variation of linear coefficient:

$$S_{N \times \beta} = \frac{(L_1 - L_2)^2}{2r} = 11.3030 \quad (f = 1).$$

Error variation:

$$S_e = S_T - S_\beta - S_{N \times \beta} = 5.1673 \quad (f = 4).$$

Table 7.2 Data of experiment no. 1 (sec)

Noise factor		Signal factor	M_1 (130 mm)	M_2 (260 mm)	M_3 (390 mm)	Linear form
N_1	Front: 70%	Rear: 30%	7.17	11.52	15.66	L_1
N_2	Front: 30%	Rear: 70%	5.23	8.83	12.17	L_2

Error variance:

$$V_e = \frac{S_e}{4} = 1.29182.$$

Noise variance:

$$V_N = \frac{S_{N \times \beta} + S_e}{5} = 3.29406.$$

S/N ratio:

$$\eta = 10 \log \frac{\frac{1}{2r}\left(S_\beta - V_e\right)}{V_N} = -33.70 \quad (db).$$

Sensitivity:

$$S = 10 \log \frac{1}{2r}\left(S_\beta - V_e\right) = -28.52 \quad (db).$$

The S/N ratios and the sensitivities found for every set of experimental conditions are given in Table 7.3. Based on those results, response graphs shown in Figure 7.6 were made.

7.8 Selection of Optimal Condition and Confirmation Test Results

The condition $A_1B_3C_2D_1E_2F_1G_1$ that increased the S/N ratio was selected as the optimal condition because we wanted to improve the smoothness of the movement of the window glass. The initial condition was $A_1B_2C_1D_1E_1F_1G_3$. In this case, the optimal condition coincided with the condition having a small degree of sensitivity. It can be expected that efficiency will also be improved as a result of the improvement of the smoothness of window glass movement by increasing the S/N ratio.

Table 7.4 shows the Prediction and the confirmation of the S/N ratio and the sensitivity at the battery voltage of 9.4 V. Figure 7.7 shows confirmation results of the operating performances. The predicted S/N ratio, the sensitivity and the gain nearly reproduced the test results.

Table 7.3   Assignment to orthogonal array and analysis results (db)

No.	A	B	e	C	D	E	F	G	S/N Ratio	Sensitivity
1	1	1	1	1	1	1	1	1	−33.70	−28.52
2	1	1	2	2	2	2	2	2	−35.92	−26.61
3	1	1	3	3	3	3	3	3	−35.10	−23.80
4	1	2	1	1	2	2	3	3	−37.39	−25.88
5	1	2	2	2	3	3	1	1	−33.85	−26.89
6	1	2	3	3	1	1	2	2	−36.96	−24.11
7	1	3	1	2	1	3	2	3	−35.84	−27.21
8	1	3	2	3	2	1	3	1	−36.70	−26.17
9	1	3	3	1	3	2	1	2	−36.71	−24.37
10	2	1	1	3	3	2	2	1	−35.02	−26.44
11	2	1	2	1	1	3	3	2	−39.37	−25.04
12	2	1	3	2	2	1	1	3	−42.32	−22.26
13	2	2	1	2	3	1	3	2	−37.46	−26.02
14	2	2	2	3	1	2	1	3	−39.53	−26.01
15	2	2	3	1	2	3	2	1	−39.60	−25.21
16	2	3	1	3	2	3	1	2	−35.15	−26.41
17	2	3	2	1	3	1	2	3	−39.44	−24.32
18	2	3	3	2	1	2	3	1	−32.14	−28.42

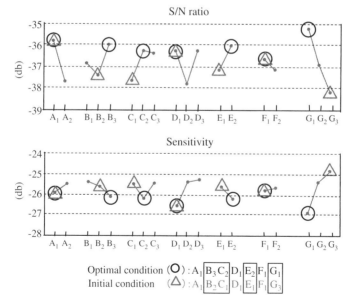

Figure 7.6   Response graphs

Table 7.4 Prediction and confirmation (battery voltage: 9.4 V) (db)

	S/N ratio		Sensitivity	
	Prediction	Confirmation	Prediction	Confirmation
Optimal condition	−31.48	−32.05	−29.37	−29.17
Initial condition	−38.50	−37.38	−25.41	−26.28
Gain	7.02	5.33	−3.96	−2.89

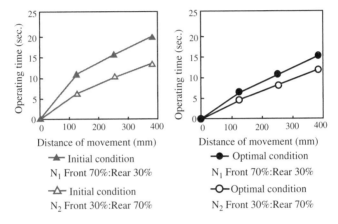

Figure 7.7 Confirmation test results for operating characteristics (battery voltage: 9.4 V)

Operating performances at an ordinary battery voltage of 13.5 V are shown in Figure 7.8. Although the degree of the improvement was smaller compared with the results obtained at 9.4 V, the variation in operating time was reduced by 31% and the operating time was reduced by 9%. In other words, even at the standard battery voltage of 13.5 V, the confirmation was made that the operating performances of the window regulator system were improved under the optimal condition compared with the initial condition.

7.9 Evaluation of Quality Characteristics

To confirm the improvement, measurements were made for the torque needed to operate the window regulator systems, which have been quality

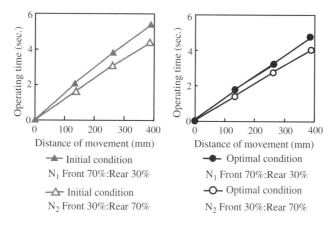

Figure 7.8   Confirmation test results for operating characteristics (battery voltage: 13.5 V)

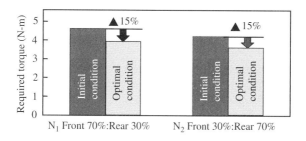

Figure 7.9   Reduction in torque required for operation (battery voltage: 13.5 V)

characteristics. As seen in Figure 7.9, the torque required for operating the window regulator system was reduced by 15% under the optimal condition compared with the initial condition. That improvement is attributed to the reduction of the sliding resistance.

Tests were also conducted to confirm the operating durability of the window regulator system. The elongation of the cable, that is, the amount of permanent deformation, was measured after the test in which the window glass was repeatedly lowered and raised for a specified number of times. Under the optimal condition, the elongation reduced by 7.3% compared with the initial condition. This result verified that durability was also improved by reducing the sliding resistance.

7.10 Concept of Analysis Based on Standardized S/N Ratio

In addition to the conventional analysis, the analysis based on the standardized S/N ratio was calculated to verify the validity of the analysis.

First, we refer to the concept of the standardized S/N ratio. Under the standard design condition N_0 with a front/rear load distribution of 50/50%, we use the output values of the L_{18} orthogonal array experiment. These values are newly adopted as the signal factor and are plotted along the horizontal axis. Accordingly, values of signal factors are all different under 18 sets of specification.

Output values under the N_1 and the N_2 conditions are then plotted along the vertical axis as shown in Figure 7.10. The standardized S/N ratio is calculated in the same way as before. Therefore, only the variation due to the N_1 and the N_2 conditions is evaluated, and the degree of nonlinearity between the distance of window glass movement and the operating time is not evaluated.

Next, we refer to how to calculate the sensitivities. As shown in Figure 7.11, the target operating times are plotted along the horizontal axis and the output values under the standard condition N_0 are plotted along the vertical axis. In this graph, the small value of the slope $ß$ expresses the mean deviation from the target value. If the value of $ß_1$ equal to 1 is obtained, it indicates that the operating time characteristic under the standard condition almost coincides with the target value. On the other hand, the value of $ß_2$ expresses nonlinearity under the standard condition N_0.

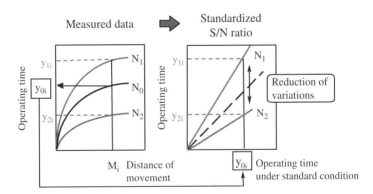

Figure 7.10 Analysis based on standardized S/N ratio: determining of standardized S/N ratio

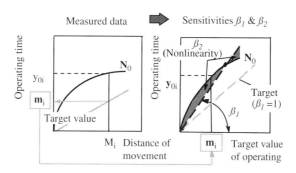

Figure 7.11 Analysis based on standardized S/N ratio: determining of sensitivities

7.11 Analysis Results Based on Standardized S/N Ratio

The data format used to calculate the standardized S/N ratio is shown in Table 7.5. Using these data, the standardized S/N ratio was calculated as follows:

Total sum of squares:

$$S_T = 7.17^2 + 11.52^2 + \cdots + 12.17^2 = 682.7856 \quad (f = 6)$$

Effective divisor:

$$S_1 = 6.38^2 + 10.74^2 + 14.73^2 = 373.0249.$$

Linear form:

$$L_1 = 6.38 \times 7.17 + 10.74 \times 11.52 + 14.73 \times 15.66 = 400.1412$$
$$L_2 = 6.38 \times 5.23 + 10.74 \times 8.83 + 14.73 \times 12.17 = 307.4657.$$

Variation of proportional term:

$$S_\beta = \frac{(L_1 + L_2)^2}{2S_1} = 671.1449 \quad (f = 1).$$

Table 7.5 Data format for determining standardized S/N ratio (data for experiment no. 1) (sec)

Signal	N_0: Standard condition	6.38	10.74	14.73	Linear form
Output	N_1: Negative side condition	7.17	11.52	15.66	L_1
	N_2: Positive side condition	5.23	8.83	12.17	L_2

Variation of linear coefficient:

$$S_{N \times \beta} = \frac{(L_1 - L_2)^2}{2S_1} = 11.51239 \quad (f = 1).$$

Error variation:

$$S_e = S_T - S_\beta - S_{N \times \beta} = 0.1284 \quad (f = 4).$$

Error variance:

$$V_e = \frac{S_e}{4} = 0.03210.$$

Noise variance:

$$V_N = \frac{S_{N \times \beta} + S_e}{5} = 2.32814.$$

Standardized S/N ratio:

$$\eta = 10 \log \frac{\dfrac{1}{2S_1}\left(S_\beta - V_e\right)}{\dfrac{V_N}{2S_1}} = 24.60 \, (db).$$

The data format used to calculate sensitivities is shown in Table 7.6. Coefficients of the first and the second terms of the sensitivity were calculated as follows.

Total variation:

$$S_T = 6.38^2 + 104^2 + 14.73^2 = 372.0249 \quad (f = 3).$$

Effective divisor:

$$S_2 = 4.96^2 + 9.92^2 + 14.88^2 = 344.4224.$$

Table 7.6 Data format for determining sensitivity based on standardized S/N ratio (data for experiment no.1) (sec)

Signal	m: Target value	4.96	9.92	14.88	Linear form
Output	N_0: Standard condition	6.38	10.74	14.73	L_1

Linear form:

$$L_1 = 4.96 \times 6.38 + 9.92 \times 10.74 + 14.88 \times 14.73 = 357.3680.$$

Let S_3 denote sum of cubes of target values and S_4 sum of fourth powers,

$$S_3 = 4.96^3 + 9.92^3 + 14.88^3 = 4392.861696$$
$$S_4 = 4.96^4 + 9.92^4 + 14.88^4 = 59313.39481088.$$

Linear form:

$$L_2 = \left(4.96^2 - 4.96 \times \frac{S_3}{S_2}\right) \times 6.38 + \left(9.92^2 - 9.92 \times \frac{S_3}{S_2}\right) \times 10.74$$
$$+ \left(14.88^2 - 14.88 \times \frac{S_3}{S_2}\right) \times 14.73 = -82.696521$$

Coefficient of linear term.

$$\beta_1 = \frac{L_1}{S_2} = 1.04.$$

Coefficient of quadratic term.

$$\beta_2 = \frac{L_2}{S_4 - \frac{S_3^2}{S_2}} = -0.025.$$

The standardized S/N ratios and the sensitivities calculated under each set of test conditions are given in Table 7.7.

7.12 Comparison between Analysis Based on Standardized S/N Ratio and Analysis Based on Conventional S/N Ratio

Response graphs were made for the standardized S/N ratio and were compared with response graphs for the conventional S/N ratio as shown in Figure 7.12. In this case it is seen that the nonlinearity of the regulator system is small and that the difference due to the N_1 and the N_2 condition is large. All of control factors display the same tendencies.

Table 7.7 Analysis results based on standardized S/N ratio

No.	A	B	e	C	D	E	F	G	Standardized S/N ratio (db)	Coefficient of linear term β_1	Coefficient of quadratic term β_2
1	1	1	1	1	1	1	1	1	24.60	1.04	−0.025
2	1	1	2	2	2	2	2	2	21.64	1.13	−0.032
3	1	1	3	3	3	3	3	3	24.42	1.70	−0.065
4	1	2	1	1	2	2	3	3	20.47	1.42	−0.068
5	1	2	2	2	3	3	1	1	24.81	1.13	−0.029
6	1	2	3	3	1	1	2	2	20.30	1.44	−0.043
7	1	3	1	2	1	3	2	3	22.10	0.98	−0.023
8	1	3	2	3	2	1	3	1	20.84	1.12	−0.018
9	1	3	3	1	3	2	1	2	21.03	1.59	−0.046
10	2	1	1	3	3	2	2	1	22.89	1.25	−0.033
11	2	1	2	1	1	3	3	2	18.10	1.18	−0.033
12	2	1	3	2	2	1	1	3	14.94	1.66	−0.073
13	2	2	1	2	3	1	3	2	20.04	1.29	−0.039
14	2	2	2	3	1	2	1	3	17.92	1.31	−0.045
15	2	2	3	1	2	3	2	1	17.70	1.44	−0.045
16	2	3	1	3	2	3	1	2	22.68	1.18	−0.032
17	2	3	2	1	3	1	2	3	17.87	1.34	−0.035
18	2	3	3	2	1	2	3	1	26.24	1.00	−0.027

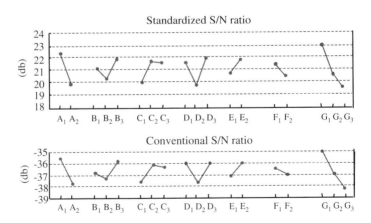

Figure 7.12 Standardized S/N ratio and conventional S/N ratio

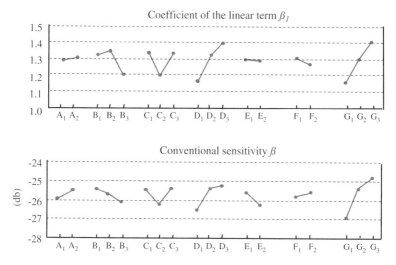

Figure 7.13 Coefficient of linear term based on standardized S/N ratio and the sensitivity based on conventional S/N ratio

Additionally, the larger effects in response graphs are seen for the standardized S/N ratio than the conventional S/N ratio.

The sensitivities were also compared in the same way. Response graphs for the coefficient β_1 of the linear term based on the standardized S/N ratio and the sensitivity based on the conventional S/N ratio are compared in Figure 7.13. For factors having a large influence, Response graphs show the same tendency as results of the conventional analysis. On the other hand, for the standardized S/N ratio of the factor like F that has small effect, the tendency differs from that obtained in the conventional analysis.

In selecting the optimal condition, levels that increase the standardized S/N ratio are selected first for the purpose of reducing variation. The levels that improve the standardized S/N ratio substantially reduce the sensitivity. Excluding the factor F, the condition here that increase the standardized S/N ratio all show desirable tendencies for the window regulator system.

Ordinarily, factors not influencing the standardized S/N ratio but having effect on the sensitivities are used to adjust to the target values, i.e. $\beta_1 = 1.00$ and $\beta_2 = 0.00$. However, in the case of the window regulator system analyzed, it is ideal that slope under the standard condition N_0 is linear and the value of β_1 is small. Therefore $\beta_1 = 1$ and $\beta_2 = 0$ are desirable.

Moreover, in this window regulator system, the value of β_1 can be adjusted by using factors other than control factors selected for this study. Accordingly, the value of β_1 was evaluated simply to see the amount of adjustment that is required when the window regulator system was applied to specific automobile model.

However, the value of β_1 under the optimal condition was 0.92. In order to make this value adjust to the coefficient 1 of the target value, one conceivable way would be to change the gear ratio of the motor drive. But the difference of the sensitivity was regarded almost the same as the target value of 1.00, so it was considered that adjustment of the sensitivity was not necessary.

With regard to the standardized S/N ratio, the factor F has little effect on the value of β_2. As a result, the optimal condition obtained by the analysis based on the standardized S/N ratio is shown in Figure 7.14. This optimal condition is the same as that selected by the analysis based on the conventional analysis.

Table 7.8 shows the prediction and the confirmation test results for both the analysis based on the standardized S/N ratio and the analysis based on the conventional S/N ratio. The data indicate that the gain of the standardized S/N ratio and the sensitivities were well reproduced. So, either analysis indicates almost the same result.

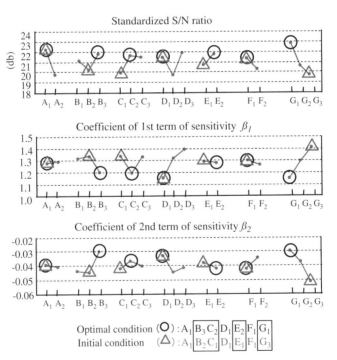

Figure 7.14   Response graphs based on standardized S/N ratio

Table 7.8 Prediction and confirmation of analysis results based on standardized S/N ratio and conventional S/N ratio

| | Analysis results based on standardized S/N ratio | | | | | |
| | S/N ratio (db) | | Coefficient of linear term β_1 | | Coefficient of quadratic term β_2 | |
	Prediction	Confirmation	Prediction	Confirmation	Prediction	Confirmation
Optimal condition	26.90	25.42	0.85	0.92	−0.015	−0.019
Initial condition	19.42	20.59	1.38	1.38	−0.053	−0.046
Gain	7.48	4.83	−0.53	−0.46	0.038	0.027

| | Analysis results based on conventional S/N ratio | | | |
| | S/N ratio (db) | | Sensitivity (db) | |
	Prediction	Confirmation	Prediction	Confirmation
Optimal condition	−31.48	−32.05	−29.37	−29.17
Initial condition	−38.50	−37.38	−25.41	−26.28
Gain	7.02	5.33	−3.96	−2.89

7.13 Conclusion

Experiments were conducted on the premise that the ideal function of a window regulator system is to have a proportional relationship between the distance of window glass movement and the operating time. The position where a weight was attached to an experimental jig was varied as the noise factor. As a result, the optimal condition was found for the window regulator system which has little sliding resistance and smooth operating characteristics. The know-how obtained in this study has already been applied to automobile productions.

The evaluations based on the S/N ratio is effective for improving experimental efficiency, making it possible to shorten development period of the window regulator system by around three months.

In the future work, we want to evaluate using a simulation instead of using of actual parts. Because the window regulator system examined in this study already had the proportional relationship and little nonlinearity, the analysis based on the standardized S/N ratio almost coincided with the analysis based on the conventional S/N ratio. Accordingly, in case of systems having a linear performance relationship, it is considered that the conventional analysis is sufficient.

7.13.1 Acknowledgments

The authors wish to express their profound appreciation to Dr. Genichi Taguchi for his valuable advices and cooperation in connection with this study.

7.14 Further Reading

1. Taguchi, Genichi (1998) *Lectures on Quality Engineering: Quality Engineering at the Development and Design Stages*, Japanese Standards Association (in Japanese).
2. Taguchi, Genichi (1999) *The Mathematics of Quality Engineering*, Japanese Standards Association, pp. 24–30 (in Japanese).
3. Taguchi, Genichi (2001) Objective functions and tuning, *Standardization and Quality Control*, **54** (8): 59–66 (in Japanese).

This case study is contributed by Takayuki Kuramochi and Shouji Miyazaki of Nissan Motor Co., Ltd, Japan and Kenji Mori of Ohi Seisakusho Co., Ltd, Japan.

8

Optimization of Next-Generation Steering System Using Computer Simulation

Nissan Motor Co., Ltd, Japan

8.1 Executive Summary

X-By-Wire systems are cutting-edge technologies in the automobile industry. Many automobile manufactures and suppliers are making a study on this new technology. However there are many technical issues to solve in order to put the technologies to practical use.

This study is concerning a next-generation steering system. In this development, achieving a natural steering feel, which customers require, is a crucial issue because the mechanism of steering-force generation in this system is fairly different from that of conventional systems. Therefore Robust Design was applied to the new steering system in order to optimize the basic design at an early phase of the development.

Robust Optimization: World's Best Practices for Developing Winning Vehicles,
First Edition. Subir Chowdhury and Shin Taguchi.
© 2016 Subir Chowdhury, Shin Taguchi, and ASI Consulting Group, LLC.
Published 2016 by John Wiley & Sons, Ltd.

Dr. Taguchi has recommended combining a computer simulation and Robust Design using standardized S/N ratio to enhance development efficiency. Corresponding to the suggestion, a computer simulation MATLAB was used to compute the steering force and standardized S/N ratio was applied to evaluate the robustness of steering force in this study. As a result, an initial goal of this development could be achieved much more effectively than expected.

8.2 Introduction

Drivers always feel cornering force and road surface conditions through a steering system while driving. These are called "Steering Information" and that is very important for drivers. Recently, advanced steering systems such as active steering system and steer-by-wire have tended to be adopted in new models or concept cars as the progress in electronics-controlled technologies.

In this study, in order to achieve an ideal steering force effectively, Robust Design was applied to the basic research on the next-generation steering system.

The optimization procedure was implemented through the following two steps in accordance with the principle of Robust Design.

The first step: Robustness of steering force against noise factors was evaluated and improved using standardized S/N ratio. Targeted values of steering force, which customers require, should be set outside the scope of consideration in this step.

The second step: Steering force was tuned to targeted values using orthogonal polynomial deployment. This step is so called "tuning" in Robust optimization. Through this tuning, steering force was tuned to targeted values almost perfectly.

This optimization procedure is the typical practice of two-step optimization that is the most essential concept of Robust Design. Through this study, optimization of the next-generation steering system could be completed in fairly short order.

8.3 System Description

The next-generation steering system studied in this chapter consists of some devices, sensors and ECU. These components should be designed to generate the appropriate steering force on a steering wheel. Therefore they were treated as control factors.

This new system is a cutting edge of technology and still under intensive research and development. Therefore its architecture is still confidential in detail.

8.4 Measurement Data

In this study, steering torque was computed by a MATLAB simulation and treated as the measurement data. On conventional steering systems, steering torque varies with steering angles and steering speeds because of its mechanism. And many drivers have been familiar to this steering feel for a very long time. Therefore the steering feel on the new system must be tuned to that of conventional one.

However the mechanism of steering-torque generation in the new system is fairly different from that of conventional one. Therefore it is difficult to achieve the required steering torque in this kind of development. Especially the relation between steering speed and steering torque is difficult to be tuned to various required values. Therefore that relation was focused on in this study.

In a MATLAB simulation, steering angles that change with elapsed time shown in Figure 8.1(a) are the input data. And torque on a steering wheel shown in Figure 8.1(b) are the output data. And the steering torque passing through a neutral position was treated as the measurement data. The torque increases in proportion to the steering speed on the neutral position because this system has damping factors in it.

Finally these data were arranged as the relation between steering speed and steering torque as shown in Figure 8.2.

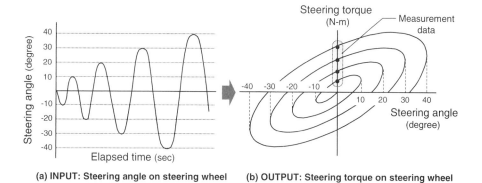

(a) INPUT: Steering angle on steering wheel   (b) OUTPUT: Steering torque on steering wheel

Figure 8.1 Measurement data

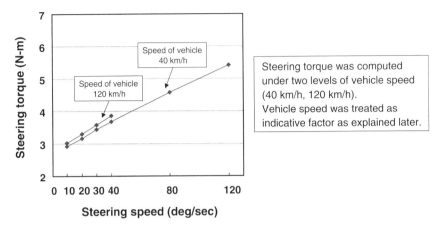

Figure 8.2   Steering speed – steering torque characteristic

8.5 Ideal Function

The steering speed – steering torque characteristic is shown in Figure 8.2. Firstly, this relation should be minimally sensitive to various noise factors. Secondly, the characteristic should be tuned to an ideal line.

The ideal line varies depending on the categories of vehicles and customers' requirements. In this study, the tuning to one of the ideal line was attempted.

8.6 Factors and Levels

The parameters in this study are given in the P-diagram shown in Figure 8.3.

8.6.1 Signal and Response

Steering speed was treated as the signal factor. In a simulation, a steering angle in the same time-interval increases gradually and the steering speed passing through the neutral position also increases. And then the steering torque just on the neutral position was treated as the response.

8.6.2 Noise Factors

Generally the slight variation in a control factor's level is treated as a noise factor in Robust optimization using a computer simulation because actual noise factors such as temperature, aging and manufacturing error finally vary

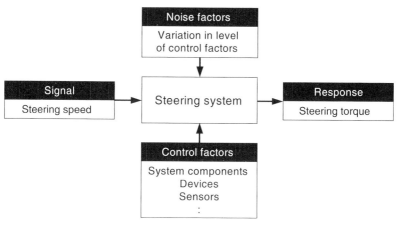

Figure 8.3 P-diagram

the control factor's level. As a result of this variation, a function of system also varies by usage conditions. Therefore, the slight variation in control factors' levels was treated as the noise factors.

8.6.3 Indicative Factor

The vehicle speed (40 km/h, 120 km/h) was treated as the indicative factor because designers can't select its optimal level and an ideal steering torque, which customers require, varies according to the vehicle speed.

8.6.4 Control Factors

Eight control factors listed in Table 8.1 were chosen from among the steering system components. And these control factors were assigned to an L_{18}.

8.7 Pre-analysis for Compounding the Noise Factors

In order to reduce the number of experiments, noise factors are often compounded into "high" and "low" conditions. Responses are maximized in "High condition" and minimized in "Low condition" due to noise effects. In this study, slight variation in the control factors' levels was treated as noises. Thus pre-simulations were conducted to know whether the level of

Table 8.1 Control factors and levels

	Control factors	Level 1	Level 2	Level 3
A	Sensor layout	1	2	—
B	Device B	1	2	3
C	Device C	1	2	3
D	Device D	1	2	3
E	Device E	1	2	3
F	Sensor sensitivity	1	2	3
G	Device G	1	2	3
H	Device H	1	2	3

steering torque increases or decreases depending on each control factor's level. The following three levels were assigned to an L_{18} orthogonal array and the average level of steering torque was computed for each run.

- Level 1: −3% of the second level of each control factor
- Level 2: the second level of each control factor
- Level 3: +3% of the second level of each control factor

Variability of control factors as noise factors were assigned to an L_{18}, and its response graph based on the pre-analysis shown in Figure 8.4. Factor A was excluded from this analysis because it was a qualitative factor. For this results, "high" and "low" noise conditions were determined with strong noise factors as shown in Table 8.2.

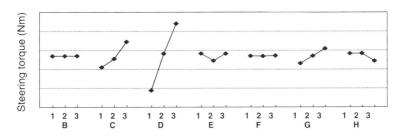

Figure 8.4 Response graph on steering torque

Table 8.2 High and low conditions

	B	C	D	E	F	G	H
N_0: Standard condition		All factors' levels are in nominal value.					
N_1: High torque condition	—	+3%	+3%	—	—	+3%	—
N_2: Low corque condition	—	−3%	−3%	—	—	−3%	—

8.8 Calculation of Standardized S/N Ratio

The data was summarized for each run of the L_{18} array as shown in Table 8.3.
   In the calculation of standardized S/N ratio, the response values under standard conditions N_0 (M_1, M_2, —M_{10}) are treated as signal factor's levels and the response values under N_1 (High) and N_2 (Low) conditions are treated as the responses as shown in Figure 8.5. This data reset is the distinctive

Table 8.3 Outer array of experiment

	Steering torque (N-m)									
	Vehicle speed: 40 km/h						Vehicle speed: 120 km/h			
Steering speed (deg/sec)	10	20	30	40	80	120	10	20	30	40
N_0: Standard condition	M_1	M_2	M_3	M_4	M_5	M_6	M_7	M_8	M_9	M_{10}
N_1: High torque condition	y_1	y_2	y_3	y_4	y_5	y_6	y_7	y_8	y_9	y_{10}
N_2: Low torque condition	y_{11}	y_{12}	y_{13}	y_{14}	y_{15}	y_{16}	y_{17}	y_{18}	y_{19}	y_{20}

 Original data   *Data reset for analysis using standardized S/N ratio*

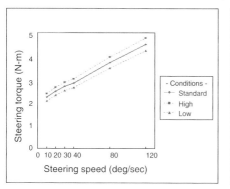
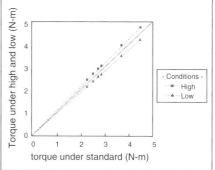

Figure 8.5 Data reset for analysis using standardized S/N ratio

feature of standardized S/N ratio. Through this data reset, pure robustness of every relation including nonlinear relations can be evaluated as zero-point proportional dynamic characteristic.

The standardized S/N ratio is calculated from the data shown in Table 8.3. Computation of S/N is as shown below.

$$S_T = y_1^2 + y_2^2 + \cdots + y_{20}^2 \quad (f = 20)$$
$$L_1 = M_1 \times y_1 + M_2 \times y_2 + \cdots + M_{10} \times y_{10}$$
$$L_2 = M_1 \times y_{11} + M_2 \times y_{12} + \cdots + M_{10} \times y_{20}$$
$$r = M_1^2 + M_2^2 + \cdots + M_{10}^2$$
$$S_\beta = \frac{(L_1 + L_2)^2}{2r} \quad (f = 1)$$
$$S_N = S_T - S_\beta \quad (f = 19)$$
$$V_N = \frac{S_N}{19}$$
$$S/N \text{ ratio} = 10 \times \log \frac{2r}{V_N}.$$

Table 8.4 Assignment of control factors and analysis results

No.	A	B	C	D	E	F	G	H	S/N ratio (db)
1	1	1	1	1	1	1	1	1	19.66
2	1	1	2	2	2	2	2	2	31.93
3	1	1	3	3	3	3	3	3	38.59
4	1	2	1	1	2	2	3	3	19.61
5	1	2	2	2	3	3	1	1	31.90
6	1	2	3	3	1	1	2	2	38.58
7	1	3	1	2	1	3	2	3	30.51
8	1	3	2	3	2	1	3	1	37.03
9	1	3	3	1	3	2	1	2	28.59
10	2	1	1	3	3	2	2	1	39.95
11	2	1	2	1	1	3	3	2	34.28
12	2	1	3	2	2	1	1	3	37.52
13	2	2	1	2	3	1	3	2	37.39
14	2	2	2	3	1	2	1	3	39.65
15	2	2	3	1	2	3	2	1	34.87
16	2	3	1	3	2	3	1	2	39.03
17	2	3	2	1	3	1	2	3	32.72
18	2	3	3	2	1	2	3	1	39.33

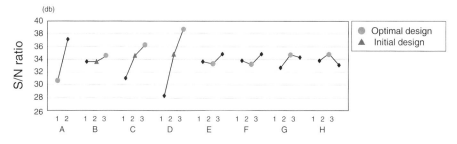

Figure 8.6 Response graph on S/N ratio

Generally a slope β or sensitivity (db) is not calculated in the analysis using standardized S/N ratio because the slope β always results in almost equal to 1.

8.9 Analysis Results

The S/N ratio for each run of the L_{18} array was calculated. The results are shown in Table 8.4 and the response graph made from these S/N ratios is shown in Figure 8.6.

8.10 Determination of Optimal Design and Confirmation

From the response graph on the S/N ratio, the optimal design for maximizing the S/N ratio was determined as follows: $A_1 \ B_3 \ C_3 \ D_3 \ E_2 \ F_2 \ G_2 \ H_2$.

In contrast, the initial design is as follows: $A_1 \ B_2 \ C_2 \ D_2 \ E_2 \ F_2 \ G_2 \ H_2$.

The second level of factor A couldn't be chosen due to special restriction. Concerning the factor E and F that showed V-shape response, the second levels were chosen for optimum. Table 8.5 shows the predicted values of S/N ratio on the optimal design and the initial one along with the confirmation results on computer simulations.

As shown in Table 8.5, good reproducibility on S/N ratio could be confirmed.

Table 8.5 Prediction and confirmation on S/N ratio

	S/N ratio	
	Prediction	Confirmation
Optimal Design	38.87 db	38.01 db
Initial Design	32.29 db	31.92 db
Gain	6.58 db	6.09 db

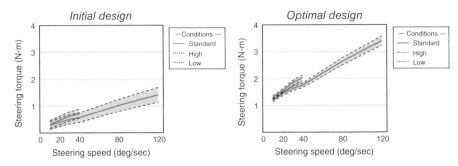

Figure 8.7   Confirmation results on initial and optimal design

The confirmed gain on S/N ratio is 6.09 db against 6.58 db of the prediction. Figure 8.7 shows input-output relations on the initial design and the optimal one. It is apparent that the optimal design has less variation of the steering torque than the initial one.

Targeted values of steering torque were set outside the scope of consideration so far. The robustness of system function is the only concern in this step.

8.11   Tuning to the Targeted Value

A tuning is conducted under a standard condition in two-step optimization procedure.

The steering torque on the optimal design under a standard condition was compared with targeted values as shown in Figure 8.8. The targeted values

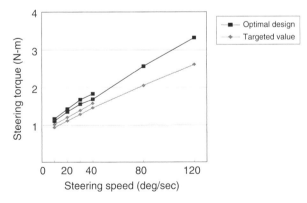

Figure 8.8   Comparison of the optimal design and targeted values

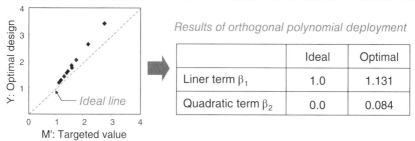

	Steering torque under standard condition (N-m)									
	Vehicle speed : 40 km/h						Vehicle speed : 120 km/h			
Steering speed (deg/sec)	10	20	30	40	80	120	10	20	30	40
M': Targeted value	M'_1	M'_2	M'_3	M'_4	M'_5	M'_6	M'_7	M'_8	M'_9	M'_{10}
Y : Optimal design	y_1	y_2	y_3	y_4	y_5	y_6	y_7	y_8	y_9	y_{10}

Results of orthogonal polynomial deployment

	Ideal	Optimal
Liner term β_1	1.0	1.131
Quadratic term β_2	0.0	0.084

Figure 8.9 Calculation of linear term β_1 and quadratic term β_2

were determined based on requests from excellent evaluaters on steering feel in NISSAN.

As shown in Figure 8.8, the steering torque on the optimal design is fairly high compared with the targeted values. To quantify this fact, the targeted values M' and the steering torque y on the optimal design under a standard condition were arranged as shown in Figure 8.9. And then linear term β_1 and quadratic term β_2 were calculated using orthogonal polynomial with treating the targeted values M' as variables.

In this analysis, as the β_1 is tuned to 1.0 and the β_2 is tuned to 0.0 more closely, the steering torque is tuned to the targeted values more accurately. The β_1 and β_2 on the optimal design are a little larger than ideal values as shown in Figure 8.9.

It is ideal that a tuning is done with control factors that affect the β_1 and β_2 but that do not affect the S/N ratio. To know each control factor's effect on the β_1 and β_2 around the optimal levels, levels of the optimal design were reset to the second level and the first and third levels were set in the vicinity of the optimal levels and then all control factors were reassigned to an L_{18} array. After this assignment, the steering torque under a standard condition was computed for each run of the L_{18} array and β_1, β_2 were calculated using orthogonal polynomial deployment.

The response graphs on the β_1 and β_2 around the optimal levels are shown in Figure 8.10. Based on this results, factor C and factor H were chosen as the tuning factors.

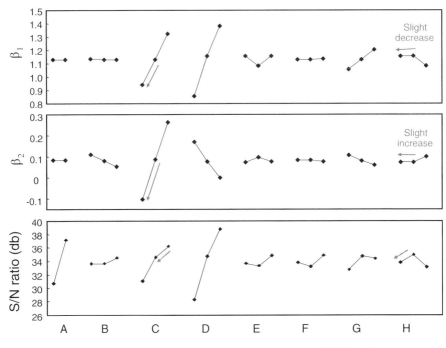

Figure 8.10   Response graphs on β_1, β_2 and S/N ratio

And the levels of factor C and factor H were decreased slightly to reduce the β_1 and β_2 on the optimal design with maintaining the higher S/N ratio as possible.

As a result of this tuning, the β_1 was improved from 1.131 to 1.004 and the β_2 was improved from 0.084 to −0.001. And the tuned optimal design was almost perfectly tuned to the targeted values as shown in Figure 8.11.

8.12 Conclusion

As a result of this typical two-step optimization practice, robustness against the noises was substantially improved. The variance of steering torque on the tuned optimal design was reduced to one-fourth compared with that of the initial one. And the levels of steering torque were almost perfectly tuned to the targeted values.

It became clear that the tuning procedure using orthogonal polynomial deployment was very effective. Using this sophisticated procedure, the

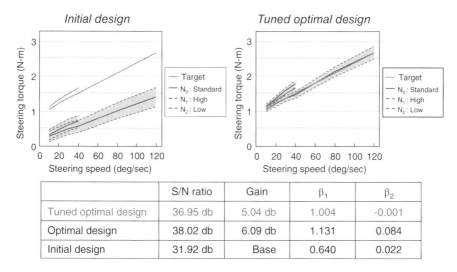

Figure 8.11 Initial design and tuned optimal design

	S/N ratio	Gain	β_1	β_2
Tuned optimal design	36.95 db	5.04 db	1.004	-0.001
Optimal design	38.02 db	6.09 db	1.131	0.084
Initial design	31.92 db	Base	0.640	0.022

steering torque can be easily tuned to targeted values that vary depending on each project. In fact, this study had been completed in only one and a half months including the planning of experiment, simulation, analysis and confirmation.

Through the success of this basic research, it was proved that applying the Robust Design to an early phase of new technology development was very effective for promoting the efficiency. And it was also proved that using computer simulations for experiments could bring us synergistic effects with applying the Robust Design. The two-step optimization procedure studied in this chapter will be extended across many kinds of engineering fields to enhance the corporate competitiveness in future.

8.12.1 Acknowledgment

We are deeply grateful to Dr. Taguchi for his considerate guidance on this study.

This case study is contributed by Kazuo Hara, Kanya Nara, Zhao Ma of Nissan Motor Co., Ltd, Japan.

9

Future Truck Steering Effort Robustness

General Motors Corporation, USA

9.1 Executive Summary

In an endeavor to improve upon historically subjective and hardware-based steering tuning development, this team was formed to find an optimal, objective solution using Design For Six Sigma. Our goal was to determine the best valve assembly design within our hydraulic power-steering assist system to yield improved steering effort and feel robustness for all vehicle models in the Future Truck program. The methodology utilized was not only multifaceted with several Design of Experiments (DOEs), but also took advantage of a CAE-based approach leveraging modeling capabilities in ADAMS for simulating full-vehicle, On-Center Handling behavior. The team investigated 13 control factors to determine which factors and settings minimized a realistic, compounded noise strategy while also considering the ideal steering effort function (SEF) desired by the customer. In the end, we learned that response-dependent variability dominated the physics of our valve assembly design concept. This meant that it would be difficult with our design and its specific set of control factors to achieve both the ideal SEF and a

Robust Optimization: World's Best Practices for Developing Winning Vehicles,
First Edition. Subir Chowdhury and Shin Taguchi.
© 2016 Subir Chowdhury, Shin Taguchi, and ASI Consulting Group, LLC.
Published 2016 by John Wiley & Sons, Ltd.

consistent SEF with minimal variation (i.e., a high Signal-to-Noise Ratio) through Parameter Design optimization only. It was also found that a certain ideal SEF was indeed desired by the customer when determined in conjunction with other key vehicle performance parameters, and that steering effort robustness, though also important, was nearly an order of magnitude lower in value to the customer in terms of dollars they would spend. Finally, we learned that designs approaching this ideal SEF could also benefit from further tolerance control. In other words, we realized that different strategic competitors tuned their valve assembly designs decidedly different – along the lines of either a high-satisfaction or a high-value strategy.

9.2 Background

Current methods for vehicle development focusing on hydraulic power-steering assist system tuning have primarily been based on iterative subjective assessments using pre-production vehicles. While a hardware-based approach such as this does find, highlight, and address many refinement issues in vehicle steering systems, it can also be costly, subjectively biased, time-consuming, and often performed too late in the development process to fully benefit a vehicle's design. And usually with only a small sample of pre-production vehicles available to evaluate, this approach cannot always comprehend the array and range of vehicle performance seen in the entire population for a given vehicle program. Thus, suboptimal steering effort tuning and potentially lower customer satisfaction can result from a hardware-based vehicle development approach.

9.2.1 Methodology

A new approach was devised for a future truck program to leverage recent advancements in Computer-Aided Engineering (CAE) modeling and simulation in conjunction with Design For Six Sigma (DFSS). Implementation of such an approach should improve steering effort tuning and optimization in two key ways. First, a CAE-based approach can objectively characterizes the physics of power-steering assist through the development of functional relationships in modeling whereas a hardware-based approach arrives at its understanding of steering system physics in a phenomenological and iterative manner. The ideal SEF desired by the customer (i.e., the driver) could thus be determined faster and more accurately via CAE methods due to better functional understanding. Second, sources of variation can be easily considered within a CAE-based approach through the use of DOEs and relatively fast simulation times whereas limited numbers of pre-production vehicles restrict such variation studies and consideration using hardware. For these reasons, a CAE-based approach should indeed produce a more optimal power-steering system design that is not only robust to variation, but also more in tune with the ideal SEF desired by the customer. These two key reasons encourage our use of a CAE-based approach for this future truck program.

Our CAE approach herein utilized the ADAMS software package along with the corporate-standard approach for modeling and simulating full-vehicle, On-Center Handling. Figure 9.1 gives an illustration of one of our full-vehicle ADAMS models. Full-vehicle modeling and simulation was necessary to completely capture the functional relationships between design parameters within the power-steering assist system, specifically focusing on the valve assembly, and the resulting SEF at the driver. The implementation of DFSS utilized Taguchi optimization techniques within several DOEs to generate our findings. Each of these optimization techniques will be defined and detailed in sections to follow.

9.2.2 Hydraulic Power-Steering Assist System

While our modeling and simulation approach did utilize full-vehicle models, our primary interest was focused on key subsystem design parameters within

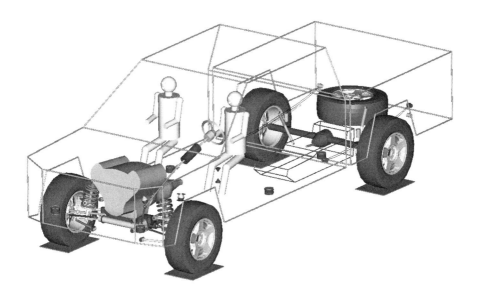

Figure 9.1   Example ADAMS model for our future truck program

the vehicle's power-steering assist system. Figure 9.2 shows an illustration of the general steering system. This type of steering system is referred to as a rack and pinion steering gear since it incorporates a rack and pinion gear set to convert steer action by the driver though the steering wheel and column into steer angle at the front road wheels. All vehicles within this future truck program utilize this same type of steering gear. The steering gear also incorporates a hydraulic power-steering assist system to counteract forces reacting back through the rack and pinion steering gear and aid the driver in steering the vehicle by managing steering effort levels. A complete structure diagram for a rack and pinion steering gear with hydraulic power-steering assist can be found in Figure 9.15 in the appendix to this chapter for reference.

A hydraulic power-steering assist system consists of five main elements. They are a hydraulic pressure circuit, a hydraulic pump, a fluid reservoir, a flow-control valve, and a piston cylinder integrated into the design of the steering gear. The pump acts to provide the hydraulic pressure in the system circuit. The fluid reservoir exists to hold excess power-steering fluid in the

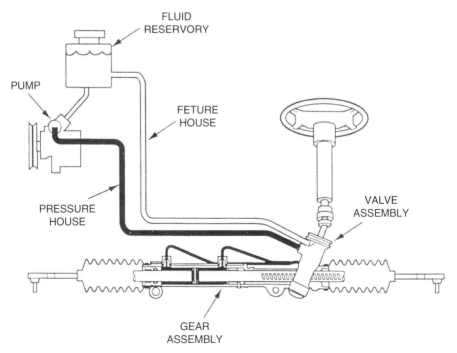

Figure 9.2 Illustration of a typical rack and pinion steering gear with power-steering assist

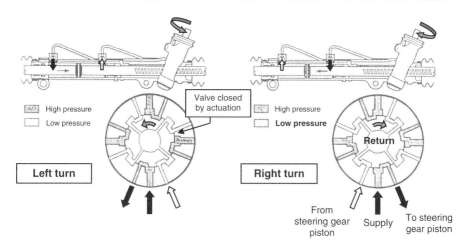

Figure 9.3 Illustration of hydraulic power-steering assist functionality for a rack and pinion steering gear

unpressurized portion of the system. The piston cylinder, or steering gear piston, provides an area for hydraulic pressure to act. And the flow-control valve is a rotary valve used to direct hydraulic pressure to either side of the steering gear piston. These elements all work together to produce steering effort assist. All of these elements can be seen in Figure 9.2.

Figure 9.3 shows the general function of a power-steering assist system during a left and right turn. As shown in Figure 9.3, hydraulic pressure generated by the steering pump is directed to either side of the steering gear piston based on the steer event and associated actuation of the flow-control valve. While driving straight down the road, the valve is effectively open producing a pressure balance on both sides of this piston. During a steer event, the valve actuates to close in one direction, causing overall system pressure to increase while also throttling hydraulic flow and decreasing pressure to the closing side. The resulting pressure differential acts on the steering gear piston in such a manner as to provide a net force to help reduce the steer input effort required by the driver to maneuver the vehicle. The amount of effort assist provided and the resulting overall SEF at the driver are not only a function of steer angle and vehicle properties, but also a function of the design of this valve. Historically, the design of the rotary, flow-control valve has been used as the primary tuning element within a power-steering assist system to alter the overall SEF to meet customer expectations for a given vehicle. Thus, any attempt to optimize the SEF at the driver must focus on the design of the valve.

9.2.3 Valve Assembly Design

The valve is typically located in series between the pinion gear and the steering column linkage as shown in Figure 9.2. The pinion gear, valve, and steering column input shaft are part of an assembly together referred to as the valve assembly. Figure 9.4 and Figure 9.5 illustrate the valve assembly and an exploded view with additional details, respectively.

The valve itself consists of a series of axial slots and intricate machined features placed on the steering gear input shaft, or spool shaft, and an outer valve body containing a corresponding set of slots and features. The valve functions through the use of a torsional centering spring in the assembly. This torsional spring, as shown in Figure 9.5, connects the spool shaft to the pinion gear and valve body. When the driver of a vehicle inputs a steer angle, the torsional centering spring deflects accordingly, creating angular separation between the slots on the spool shaft and the slots on the valve body. This angular separation defines the actuation of the valve and the direction of hydraulic pressure and flow. The effort feedback felt by the driver for this steer input results from the difference between the mechanical system response (i.e., the deflection torque of the torsional spring and the net cornering force acting back through the rack and pinion steering gear) and the hydraulic assist force at the steering gear piston.

As shown in these figures and from this explanation, the valve assembly is indeed a complex design with many parameters. Determining the scope and

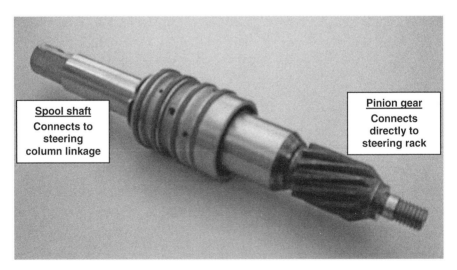

Figure 9.4   Valve assembly design for our future truck program

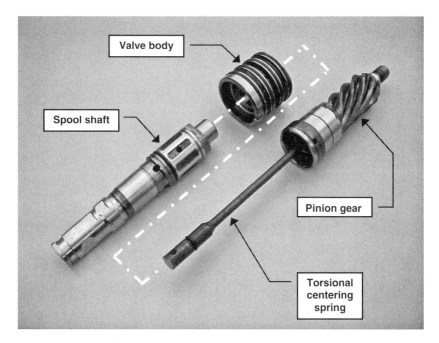

Figure 9.5 Exploded view of a typical valve assembly

influence of these design parameters in optimizing the SEF will be the focus of this project.

9.2.4 Project Scope

The scope of this study was primarily limited to the design of the valve assembly along with a few key parameters in the power-steering assist system. Other factors were included to understand their relative impact on steering effort tuning and optimization, but were not part of our design recommendations.

Furthermore, this study was also limited to the steering effort tuning and optimization for the high-volume vehicle models. Even though just a subset of the program was directly studied, it is believed that the findings from this study will also be applicable to all similar vehicle models.

Finally, this project focused only on steering effort tuning in relation to On-Center Handling performance. Steering efforts in other performance areas were considered but not directly studied in this project.

9.3 Parameter Design

9.3.1 Ideal Steering Effort Function

The first step in Parameter Design optimization was to determine the ideal SEF desired by the customer. This step involved choosing the best function to represent the SEF based on engineering judgment and then comparing a wide selection of vehicles to determine the most desirable or ideal SEF.

Through our CAE-based approach, we were able to understand the functional relationship between driver steering input and steering effort response in terms of the dominate elements in the vehicle steering system. Figure 9.6 illustrates that functional relationship. As shown in Figure 9.6, the relationship can be represented as a control system loop having six dominate elements, including the driver. The benefit of having such a system is that we are free to choose any point within the loop to measure the system and perform optimization. Lateral acceleration was chosen instead of steering wheel angle as our independent variable based on improved ease of comparison with established vehicle performance metrics. And since steering efforts were of primary importance, the best SEF for our project was determined to be steering wheel torque as a function of lateral acceleration. This choice of SEF is

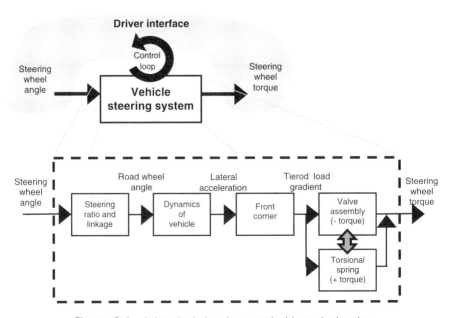

Figure 9.6 Intended steering control loop behavior

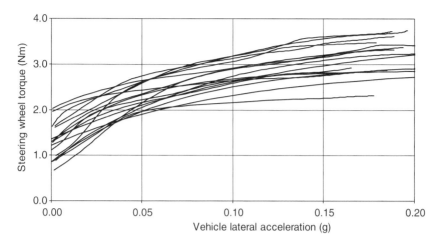

Figure 9.7 Comparison of SEFs from several vehicles from different manufacturers

often used by others in the automotive industry when it comes to steering effort tuning and comparison [1].

Figure 9.7 provides a sample of the competitive assessment we conducted to determine the ideal SEF desired by the customer in terms of steering wheel torque versus lateral acceleration. Our analytical approach allowed us to numerically compare the SEFs of many vehicles from different manufacturers. Taken together with customer feedback and expert opinion for steering effort and feel satisfaction, it was found that one SEF was indeed the most desirable. This particular SEF was from a vehicle and a strategic competitor known for desirable steering performance, and this vehicle will be referred to as Vehicle A. The SEF for Vehicle A is shown in Figure 9.8. For the purposes of our Parameter Design optimization, the SEF from Vehicle A was used as our ideal SEF. Figure 9.8 also includes a second SEF from a vehicle herein referred to as Vehicle B. The SEFs of these two vehicles differ significantly with Vehicle B's SEF being more nonlinear with respect to lateral acceleration and having a higher initial steering wheel torque than Vehicle A's SEF. Additionally, the satisfaction level associated with the SEF of Vehicle B was considerably reduced in terms of both customer feedback data and internal expert opinion. Vehicle B was from strategic competitor known for high quality and is also a direct competitor to our future truck program. For these reasons, the team decided to include Vehicle B's SEF in the Parameter Design optimization to study it as an alternate SEF.

Optimization was performed around each of the two chosen SEFs and details of the findings are provided in the following sections.

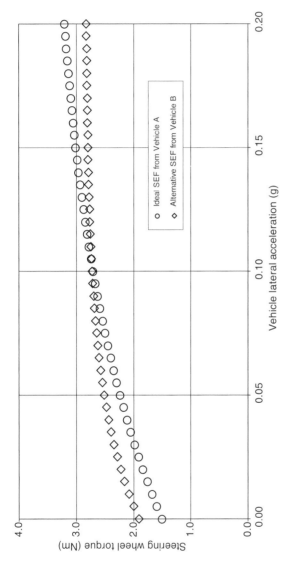

Figure 9.8 Two chosen SEFs for parameter design optimization

9.3.2 Control Factors

With both the ideal and alternative SEF targets established, the next step was to define the details for the Parameter Design DOE. Thirteen control factors in total were selected and studied in this DOE. The rationale for selecting these 13 factors was multifaceted. Several factors were selected based on their historic use as steering effort tuning parameters. They comprised features of the valve assembly design related to either the torsional centering spring or the slots on the spool shaft and valve body. Additionally, these seven parameters are generally inexpensive to alter. Conversely, another set of control factors were selected based on their lack of prior usage. They comprise features of the valve assembly design that are either difficult to change once the steering architecture is set or are relatively more expensive to change. Four additional factors outside of the valve assembly were included in the DOE from parameters either within the power-steering system or from other vehicle properties. They were considered in our Parameter Design optimization because they were not only inexpensive to change, but also proved significant in previous development work for steering effort tuning.

Table 9.1 gives the level strategy used in our DOE for all thirteen control factors. As shown in Table 9.1, all factors except one were studied at three levels. Performing the DOE with factors at more than two levels enables improved insight into the nonlinear effects each control factor had on the SEF. Control Factor A was studied, however, at six levels due to its past significance and need for further insight into this parameter.

A modified orthogonal L36 array was utilized for our Parameter Design DOE. This array was appropriate for our DOE since it provided us the ability to study a large number of factors at more than two levels. The inner array was modified through column combining techniques to incorporate the Control Factor A at six levels and still maintain an orthogonal and balanced DOE. The specific format and detail for our modified L36 inner array containing all control factors and their associated level settings can be found in Table 9.3. Further information into the details, usage, and explanation of an L36 array along with its modified form can be found in various publications for optimization experiments (e.g., [2]).

9.3.3 Noise Compounding Strategy and Input Signals

To complete the setup of the DOE, it was necessary to determine the format of our outer array. The main goal of any Parameter Design optimization is to find a design that is insensitive or robust to variation in the system (i.e., noise). Our goal here was to include sources of variation in the outer array such as the

Table 9.1 Control factors and associated
levels for parameter design DOE

Control factor	Number of levels
Control factor A	6
Control factor B	3
Control factor C	3
Control factor D	3
Control factor E	3
Control factor F	3
Control factor G	3
Control factor H	3
Control factor I	3
Control factor J	3
Control factor K	3
Control factor L	3
Control factor M	3

effect of different vehicle models, the range of vehicle masses, and the use of multiple tires with different properties to find a valve assembly design that produced the most robust ideal SEF. To do so, we needed to include these sources of variation in the outer array of our L36. In an effort to simplify our DOE and reduce the number of runs, a noise compounding strategy was implemented for the outer array. This compounding strategy is detailed in Table 9.2.

Three levels for compounded noise were included in the DOE. Sources of variation were selected such that they produce either a low, nominal, or high SEF response. The noise conditions N1 and N2 provided the contrast in the study to perform optimization and noise condition N0 was necessary for the Standardized S/N post-processing utilized in this study.

Two levels of an additional input signal were included in the outer array also. These two levels D1 and D2 represented the compounded variation effect of different driving behaviors within our DOE. Along with lateral acceleration as the primary input signal, this additional input signal was included in our DOE to further add contrast and to potentially provide more insight into valve assembly design robustness. The specific format and detail for our L36 outer array containing both noise and input signals can be found

Table 9.2 Summary of compounded noise strategy and additional input signal in the outer array

Noise condition	Details
N0	Nominal response noise condition
N1	Low response, compounded noise condition based on spool valve system noise
N2	High response, compounded noise condition based on spool valve system noise
Input signal	**Details**
D1	Low response, compounded noise condition based on driver behavior
D2	High response, compounded noise condition based to driver behavior

in Table 9.4. Including these three noise and two input signal levels, a total of 216 runs were performed in this L36 Parameter Design DOE.

For a complete summary of the setup of the entire Parameter Design optimization, refer to the P-Diagram in Figure 9.16 in the appendix to this chapter.

9.3.4 Standardized S/N Post-Processing

The results from our DOE can be seen in Figure 9.9, Figure 9.10, and Figure 9.11. These results are in the form of response graphs for the S/N Ratio, Beta1, and Beta2 responses, respectively. Since our chosen SEFs were

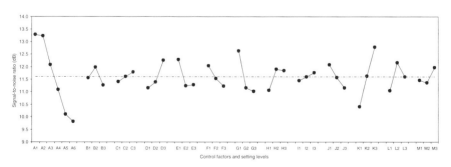

Figure 9.9 Response graph for signal-to-noise (S/N) ratio

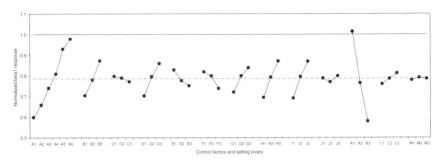

Figure 9.10   Response graph for β_1 response

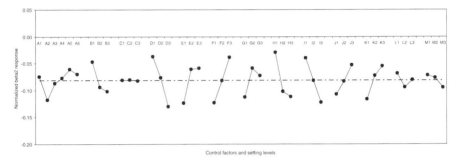

Figure 9.11   Response graph for β_2 response

nonlinear functions, a nonlinear optimization post-processing method called Standardized S/N was utilized. This style of post-processing is another Taguchi optimization technique and it produces not only the S/N Ratio response as a measure of robustness, but also not one but two beta coefficient terms to define the mean response. The Beta1 response represents the overall gain for our SEF while the Beta2 response represents the curvature for our SEF. Standardized S/N requires the consideration of these three responses together to perform optimization.

By examining the response graph for S/N Ratio in Figure 9.9, it can be seen that control factors A, G, and K have the most significant impact on S/N Ratio. For instance, a valve assembly design with control factor A set to level 1, control factor G set to level 1, and control factor K set to level 3 would be more robust than a design with A set to level 6, G set to level 3, and K set to level 1. Also, it can be seen that control factors C and I have a minimal impact on S/N Ratio. Given this response graph in Figure 9.9, the most optimal design would be to set each control factor appropriately so as to maximize S/N Ratio and produce an SEF that is highly consistent.

Our 13 control factors gave different results in the response graph for Beta1, seen in Figure 9.10. Here, the goal was to produce a Beta1 response equal to the value of one (Beta1 = 1 represents the same gain as the ideal SEF). As it can be seen, control factor A and K have the most significant impact on Beta1 response. Also, the dotted line in the center of the response graph represents the average Beta1 value of the entire DOE. The average Beta1 value was 0.79. This value was lower than the value of one shown by the solid line in the response graph, and made it necessary to use several control factors to produce an SEF similar to the ideal SEF. However, if we were to use control factor K set to level 1 for example, this would significantly reduce the S/N Ratio of the design. Moreover, if we were to use control factor A at level 6 in our endeavor to match the ideal SEF, this would also work contrary to our optimization goal of a higher S/N Ratio. By comparing Figure 9.9 and Figure 9.10, it becomes apparent that a majority of the control factors that increase robustness in our valve assembly design unfortunately adversely affect the gain response of our SEF.

With the addition of the results shown in Figure 9.11 for the Beta2 response, the conflict in selecting the parameter settings worsens. Here, the goal was to produce a Beta2 response equal to the value of zero (Beta2 = 0 represents the same curvature as the ideal SEF). As can be seen in Figure 9.11, the average DOE response for Beta2 was −0.081, a value lower than the desired value of zero to match the ideal SEF. Again, several of the 13 control factors were allocated to adjust Beta2 value to match that of the ideal SEF. Factors like D, F, H, and I were utilized to do so; however, this again reduced the S/N Ratio and therefore increased our valve assembly's sensitivity to variation yielding an inconsistent SEF in the presence of noise.

After many attempts trying to balance the opposing responses for Beta1, Beta2, and S/N Ratio, it became obvious that it is not possible using our valve assembly and its associated design parameters to achieve a SEF which is both ideal and robust to variation through Parameter Design optimization only (i.e., we can either achieve our robustness target or our ideal SEF, but not both simultaneously). This discovered property of our valve assembly design was referred to as response-dependent variability. Further investigations beyond the current design space in our DOE (not included in this chapter) have determined that this response-dependent variability is inherent in the design and physics of a rotary flow-control valve and cannot be de-coupled through axiomatic design techniques or by DOE methods.

Another way to utilize the results from Figure 9.9, Figure 9.10, and Figure 9.11 is in conjunction with the Principle of Additivity. This principle provided us with a means to interpolate the S/N Ratio performance of valve assembly designs not originally included as a part of our fractionized Parameter Design DOE. The equation for S/N Ratio utilizing the Principle of Additivity is as follows:

$$\text{S/N Ratio} = \overline{T} + \sum_{M}^{A} [(\text{Control Factor } i \text{ at desired level}) - \overline{T}]. \qquad (9.1)$$

The term $\overline{T}$ represents the average S/N Ratio response of 11.60 dB from the DOE. Similar equations to (9.1) can be developed and implemented for the Beta1 and Beta2 responses also using the appropriate $\overline{T}$. Various design confirmation studies on our part (not included in this chapter) proved that this Principle of Additivity holds for our DOE results. This meant that we could use this principle to generate valve assembly design alternatives and predict their associated S/N Ratio, Beta1, and Beta2 responses using Equation (9.1) or similar equations.

Since the Principle of Additivity was proven to hold, it was appropriate to investigate valve design alternatives that matched the SEFs for Vehicle A and B due to their importance established earlier. In doing so, we hoped to learn and understand the potential trade-offs of each valve assembly design used to produce each SEF, given our knowledge of response-dependent variability.

Figure 9.12 gives the performance of the valve assembly design alternative matching Vehicle A's SEF. Compounded noise and input signal variation conditions were also input into the specific simulation with this valve assembly alternative to understand the impact of variation on this SEF. Four variation states in total were considered, for example the combination pairs of N1, N2, D1, and D2, and the resulting effects on the SEF are shown in Figure 9.12.

As can be seen in Figure 9.12, the matching SEF experienced significant variation in response when exposed to the four variation conditions. The overall S/N Ratio of this design matching the ideal SEF was 7.47 dB, 4 dB lower than the average for the DOE. The majority of the variation in the SEF response was attributed to N1 and N2. As detailed earlier, these two conditions represented compounded noise in the system ultimately related to manufacturing variation. The effect of D1 and D2 (i.e., the compounded driver behavior variation) was much smaller than that from the noise variation. While the robustness performance of this design overall does

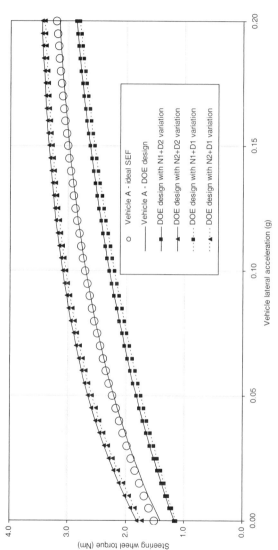

Figure 9.12 DOE design for vehicle A under noise conditions

not meet our target, it was interesting to find that the design was inherently more robust to driver behavior. It was further inferred that if this design matching the ideal SEF were implemented with better tolerance controls in manufacturing, then this would indeed be perceived as a highly desirable design by the customer. Thus, our finding of response-dependent variability required tolerance control and its associated higher costs if a manufacturer were to produce a valve assembly design matching the ideal SEF.

In comparison to the results for Vehicle A shown in Figure 9.12, a similar investigation was performed for Vehicle B and its SEF. Again using the Principle of Additivity, a valve assembly alternative was generated to closely match the SEF of Vehicle B. Next, our compounded noise and input signal variation conditions were evaluated and plotted together with this matching SEF in Figure 9.13. As shown in Figure 9.13, the overall level of robustness for this design was higher than the design matching Vehicle A. This is demonstrated by the relative closeness of all the SEF curves. The S/N Ratio of the valve assembly design matching Vehicle B's SEF was 14.60 dB. This is approximately 3 dB greater than the average for the DOE. Also, this result was 7 dB more robust or 60% less variable than Vehicle A's SEF while under the same conditions.

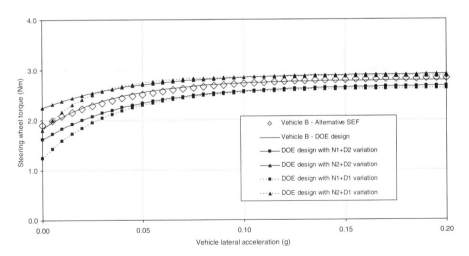

Figure 9.13   DOE design for vehicle B under noise conditions

Further investigation into Figure 9.13 shows another interesting finding. While the overall level of robustness is higher matching Vehicle B's valve assembly design, its sensitivity to the different sources of variation changed. Where Vehicle A's SEF was more sensitive to noise variation from manufacturing, the valve assembly design matching Vehicle B's SEF was more sensitive to driver behavior. This is especially true at low levels of lateral acceleration. In general, Vehicle B's SEF was not perceived well by customers or by internal experts despite its high robustness. Thus, improving consistency around a less desired SEF may not be the goal of Vehicle B's manufacturer; instead we have concluded that this valve assembly design affords its manufacturer the ability to reduce tolerance controls and tolerance control-related costs. In doing so, the likely goal of this SEF becomes less about customer satisfaction through steering performance and more about customer satisfaction through improved value.

9.3.5 Quality Loss Function

To ultimately understand the best balance between robustness and ideal response given response-dependent variability, the last step in this Parameter Design optimization was to establish a Quality Loss Function (QLF) to measure and compare valve assembly designs based on customer preference. As we have already found with response-dependent variability inherent in our valve assembly design, we cannot achieve both our targeted level of robustness and the ideal SEF by parameter setting alone. Our QLF helped us to determine the relative importance in the trade-offs between mean response and variability in terms of dollar value to the customer. That is, the QLF provided a decision-making value function that compared high-robustness valve assembly designs to ideal SEF valve assembly designs and many designs in between in terms of dollars a potential customer was willing to pay. From here, we were able to arrive at a design that indeed balanced these competing performances and provided the best net gain for our future truck program.

To establish our QLF, a similar SEF comparison was conducted with a wide selection of different vehicles as performed earlier. The difference this time was that we functionally regressed these SEFs to selected customer feedback data from J.D. Power metrics associated with steering performance that had already been linked to a dollar value at the customer. Initially, the SEF data from the selection of vehicles did not produce a regression with a high correlation (i.e., R^2 value close to 1). Several statistical techniques were used in conjunction with key additional vehicle performance parameters to produce a high-correlation regression between customer dollar value and these key vehicle performance parameters including the SEF. This final regression achieved a $R^2 = 0.924$. It is shown in Figure 9.14. This regression benefits from previous learnings regarding an internal steering performance metric referred to as "Road Feel." Also, several post-assessments of this regression were conducted to check its validity. For example, one such assessment was to determine where Vehicle A ranked in the regression. The final regression passed this assessment with Vehicle A ranked as the highest relative dollar value at the customer. It was interesting to note that Vehicle B, a low dollar value at the customer as expected, actually outperformed the regression. The team believed this was partly due to this Vehicle B's pre-existing reputation for high quality and thus the perception of good steering performance.

With the regression established for the QLF, valve assembly designs could now be compared in term of relative dollars given the set of key vehicle performance parameters for our future truck program. Using this regression, several important findings were discovered. First, it was found that as valve assembly designs produced SEFs closer to Beta1 = 1 and Beta2 = 0, they increased in dollar value approaching the value of Vehicle A. Next, it was found that as designs produced SEFs higher in S/N Ratio like Vehicle B, they too increased in dollar value, although not as much as having Beta1 and Beta2 set to ideal. In the end, the response-dependent variability and its trade-offs between mean response and variability forced the optimal valve assembly design to be quite close to the ideal SEF with considerable sensitivity to variation. This result implies that the dollar value improvement for the mean response was more valuable than the loss in value due to increased variation sensitivity. Our findings suggest this difference to be nearly an order of magnitude.

The final valve assembly design recommended for this future truck program utilized our established QLF to produce the appropriate balance between robustness and ideal SEF response. Since the QLF showed that the ideal SEF was nearly an order of magnitude higher in dollar value at the customer than the value of variation reduction, the design proposed was quite similar to the ideal SEF. The proposed design does show a relatively low S/N Ratio. At this point, specific tolerance controls could be implemented based on our Tolerance Design optimization work (not included in this chapter) to further limit QLF loss and improve customer satisfaction. The associated costs with these tolerance controls would also have to be considered in conjunction with the QLF to determine if the net gain to this future truck program could be further increased.

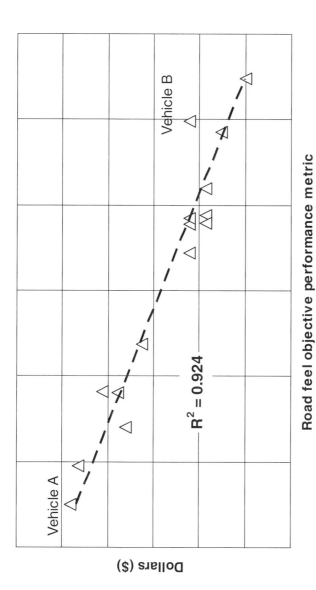

Figure 9.14 Relationship between road feel objective performance metric and value to the customer

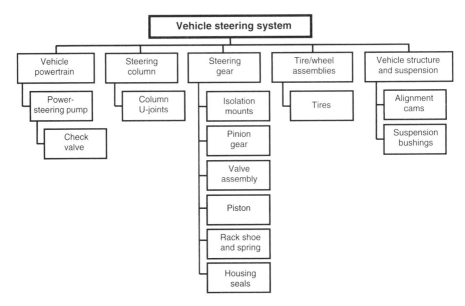

Figure 9.15   Structure diagram for our future truck steering system

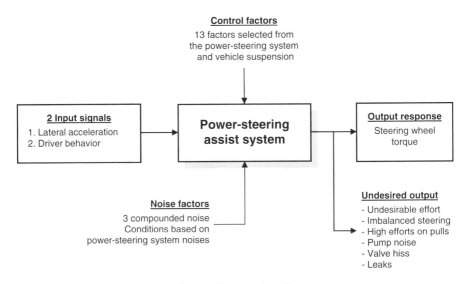

Figure 9.16   Parameter diagram

Table 9.3 Modified L36 inner array with control factor level settings implemented in parameter design DOE

Run	Control factors												
	A	B	C	D	E	F	G	H	I	J	K	L	M
1	1	1	1	1	1	1	1	1	1	1	1	1	1
2	1	2	2	2	2	2	2	2	2	2	2	2	2
3	1	3	3	3	3	3	3	3	3	3	3	3	3
4	1	1	1	1	1	2	2	2	2	3	3	3	3
5	1	2	2	2	2	3	3	3	3	1	1	1	1
6	1	3	3	3	3	1	1	1	1	2	2	2	2
7	2	1	1	2	3	1	2	3	3	1	2	2	3
8	2	2	2	3	1	2	3	1	1	2	3	3	1
9	2	3	3	1	2	3	1	2	2	3	1	1	2
10	2	1	1	3	2	1	3	2	3	2	1	3	2
11	2	2	2	1	3	2	1	3	1	3	2	1	3
12	2	3	3	2	1	3	2	1	2	1	3	2	1
13	3	1	2	3	1	3	2	1	3	3	2	1	2
14	3	2	3	1	2	1	3	2	1	1	3	2	3
15	3	3	1	2	3	2	1	3	2	2	1	3	1
16	3	1	2	3	2	1	1	3	2	3	3	2	1
17	3	2	3	1	3	2	2	1	3	1	1	3	2
18	3	3	1	2	1	3	3	2	1	2	2	1	3
19	4	1	2	1	3	3	3	1	2	2	1	2	3
20	4	2	3	2	1	1	1	2	3	3	2	3	1
21	4	3	1	3	2	2	2	3	1	1	3	1	2
22	4	1	2	2	3	3	1	2	1	1	3	3	2
23	4	2	3	3	1	1	2	3	2	2	1	1	3
24	4	3	1	1	2	2	3	1	3	3	2	2	1
25	5	1	3	2	1	2	3	3	1	3	1	2	2
26	5	2	1	3	2	3	1	1	2	1	2	3	3
27	5	3	2	1	3	1	2	2	3	2	3	1	1
28	5	1	3	2	2	2	1	1	3	2	3	1	3
29	5	2	1	3	3	3	2	2	1	3	1	2	1
30	5	3	2	1	1	1	3	3	2	1	2	3	2

Table 9.3 (*Continued*)

Run	Control factors												
	A	B	C	D	E	F	G	H	I	J	K	L	M
31	6	1	3	3	3	2	3	2	2	1	2	1	1
32	6	2	1	1	1	3	1	3	3	2	3	2	2
33	6	3	2	2	2	1	2	1	1	3	1	3	3
34	6	1	3	1	2	3	2	3	1	2	2	3	1
35	6	2	1	2	3	1	3	1	2	3	3	1	2
36	6	3	2	3	1	2	1	2	3	1	1	2	3

Table 9.4 L36 outer array for compounded noise conditions and input signals implemented in parameter design DOE

Run	D1			D2		
	N1	N0	N2	N1	N0	N2
1						
2						
3						
4						
5						
6						
7						
8						
9						
10						
11						
12						
13						
14						
15						
16						
17						
18						
19						
20						
21						
22						
23						
24						
25						
26						
27						
28						
29						
30						
31						
32						
33						
34						
35						
36						

Engineering metric:
Steering wheel torque vs. vehicle lateral acceleration

9.4 Acknowledgments

We the authors of this chapter would like to acknowledge the dedication and focus of the entire team in planning, developing, and executing this project. Without the team's dedication and focus, this project would not have been a success. Key team members in this effort were George Doerr and Mike Jones. We thank them for their efforts.

We would also like to thank our Engineering Group Manager, Max Farhad, for his foresight in letting us pursue this work and for his awareness of the value such work could bring to this future truck program and to General Motors Corporation.

We would also like to thank Jay Eleswarpu of ASI Consulting Group, LLC for his assistance with our implementation of the Standardized S/N postprocessing, which was used to produce our Parameter Design findings. His help and knowledge of Taguchi optimization techniques was greatly appreciated by the team.

9.5 References

1. Cowan, Brian (2006) More than a feeling. *Vehicle Dynamics International* **1**: 30–2.
2. Taguchi, G. and Konishi, S. (1987) *Orthogonal Arrays and Linear Graphs: Tools for Quality Engineering*. Allen Park, Michigan: ASI Press.

This case study is contributed by Jason P. Delor, and Jason Wong, of General Motors Corporation, USA.

10

Optimal Design of Engine Mounting System Based on Quality Engineering

Mazda Motor Corporation, Japan

10.1 Executive Summary

The engine mounting system of a motorcar has the function of attenuating the vibration transfer from the engine to the body and improving ride comfort. The authors applied the standard S/N ratio Taguchi Method to the design of the three-point mounting system. The design parameters are optimized so that the dispersion of the vibration spectrum estimated by the simulation becomes small so that the spectrum agrees with its target. Although the performance target has not been achieved by the usual application of Taguchi Method, by applying Taguchi Method iteratively, a mounting system with excellent robust performance has been derived. The S/N ratio has been improved by 13 dB, which means that the defective incidence has been

Robust Optimization: World's Best Practices for Developing Winning Vehicles,
First Edition. Subir Chowdhury and Shin Taguchi.
© 2016 Subir Chowdhury, Shin Taguchi, and ASI Consulting Group, LLC.
Published 2016 by John Wiley & Sons, Ltd.

reduced to 1/20. The spectrum has come to agree with the target function almost perfectly and the main peaks of the vibration are fully attenuated.

10.2 Background

The engine mount of a motorcar is a rubber element placed between the engine and the body and has the function of fixing the engine on the body and at the same time improving ride comfort by preventing the transfer of engine vibration to the body.

  In recent years, the demand of vehicle vibration and noise performance has become more serious and a mounting system with better vibration isolation performance has been needed. However, in a mounting system, the relation of the performance and design parameters is complicated and the derivation of the condition to achieve optimal isolation performance is difficult. For example, when vibration at a certain frequency is reduced, the vibration at another frequency often gets worse, or performance varies at mass production stage even if performance at development stage is good and stable. In this chapter, we try to apply quality engineering to engine mounting system design to solve this problem.

10.3 Design Object

One of the most typical engine mounting systems is the four-point system in which four mount elements are used for one vehicle. However, in recent years a three-point mounting system has become more typical for its low cost and high occupant safety to crash accidents. However, the three-point system has more disadvantages than the four-point system for vibration isolation performance, and improving it is required. In the three-point system, each mount is arranged as shown in Figure 10.1. The one end of each mount is combined with the engine through the bracket and the other end is combined with the body directly. Each mount is referred by the number #1, #3 and #4.

  Though a motorcar has a lot of vibration source, the largest demand to the mounting system is the reduction of low frequency engine vibration below 30 Hz. Its source is the variation of the engine torque around the crankshaft as shown in Figure 10.2. The magnitude of the vibration is evaluated from the spectrum of the Z direction (vertical) vibration of the vehicle floor under the driver's seat.

  As previously mentioned, in the mounting system design, it is desirable to reduce the dispersion of the isolation performance at the mass production stage

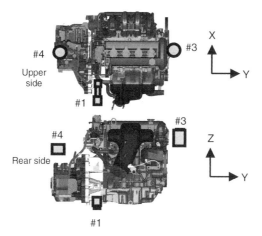

Figure 10.1   Three-point engine mounting system

and improve isolation performance itself at the same time. Therefore, the primary purpose of this chapter is to reduce performance variation due to manufacturing error, and the secondary purpose is to make the average vibration property approach its target. Therefore in this chapter, quality engineering is applied to the design of a three-point mounting system to improve isolation performance for the above-mentioned low frequency vibration.

10.4   Application of Standard S/N Ratio Taguchi Method

The main design parameters for mounting system, namely, the control factors of the isolation performance, are the spring constant, the damping coefficient

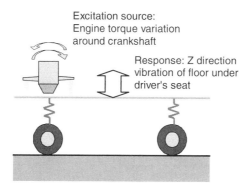

Figure 10.2   Low frequency engine vibration

and the installation position of each mount. Among these parameters, the control factor that greatly influences the low frequency vibration in particular are the spring constant and damping coefficient in the X and Z direction of each mount, and the position in the X, Y and Z direction of # 3 and #4 mount. Table 10.1 shows these control factors. In the conventional mounting system design, the value of these control factors is corrected by the empirical rule while estimating the magnitude of the body vibration by CAE simulation.

In this chapter, each control factor is optimized by applying the standard S/N ratio Taguchi Method to this vibration simulation. In the standard S/N ratio Taguchi Method, the value of the control factors is determined so that the dispersion of the performance property caused by the noise factor will be minimized, namely, the S/N ratio, the index of the smallness of the dispersion of the performance property, will be maximized. Then, the value of the control factors are tuned so that the average performance property agrees with its

Table 10.1 Control factors

Control factor	Level		
	1	2	3
#1-X spring constant	×0.7	×1	×1.3
#1-Z spring constant	×0.7	×1	×1.3
#3-X spring constant	×0.7	×1	×1.3
#3-Z spring constant	×0.7	×1	×1.3
#4-X spring constant	×0.7	×1	×1.3
#4-Z spring constant	×0.7	×1	×1.3
#1-X damping coef.	×0.4	×1	×1.6
#1-Z damping coef.	×0.4	×1	×1.6
#3-X damping coef.	×0.4	×1	×1.6
#3-Z damping coef.	×0.4	×1	×1.6
#4-X damping coef.	×0.4	×1	×1.6
#4-Z damping coef.	×0.4	×1	×1.6
#3-X position	−30 mm	+0 mm	+30 mm
#3-Y position	−30 mm	+0 mm	+30 mm
#3-Z position	−30 mm	+0 mm	+30 mm
#4-X position	−30 mm	+0 mm	+30 mm
#4-Y position	−30 mm	+0 mm	+30 mm
#4-Z position	−30 mm	+0 mm	+30 mm

target. In other words, the performance property is made to fit the target curve through the procedure in that the performance property is redrawn using the target curve as the independent variable, and the value of the control factor is chosen so that the first order slope $\beta 1$ becomes 1 and the second order slope $\beta 2$ to become 0. The optimal robust design for the noise and ideal achievable performance is derived through this two-step design.

In this chapter, the value of each control factor is determined so that the vibration spectrum under the noise condition predicted by the simulation has the minimum variation and so that the nominal spectrum agrees to the target curve; in other words, it will have an excellent transcript to the target property.

The level interval of each control factor was set to about five times the maximum manufacturing error. The spring constant has three levels, the nominal value (1.0 time), 0.7 times and 1.3 times the nominal value. The damping coefficient also has three levels, the nominal value, 0.4 times and 1.6 times the nominal value. As for the installation position, three levels are the nominal level value (+0 mm), −30 mm and +30 mm displacement from the nominal position. The noise factors are assumed to be the change of all control factors, and the analysis using the orthogonal array of L54 (control factor) × L54 (noise factor) is conducted.

Figure 10.3 shows the result of the analysis. The broken line is the original vibration spectrum and the solid line is the optimized one. The target property is the dotted line. The peak value of the vibration spectrum has been decreased only by 2 dB after the control factors are optimized, and the excellent transcript of the vibration spectrum to the target property has

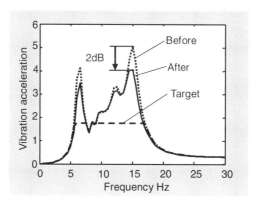

Figure 10.3   Vibration spectrum before and after optimization

not been achieved. This has not been improved even after changing the control factors and running the analysis again.

10.5 Iterative Application of Standard S/N Ratio Taguchi Method

As the reason why a good result has not been obtained in the analysis of the previous section, it is considered that in the mounting system the interaction between the control factors is strong, and the optimal solution has not been obtained. For an explanation of this, Figure 10.4 shows the concept of the relation between the control factors and the S/N ratio or the sensitivity when the interaction between the control factors is strong. If the value of the first control factor is at the point of "Initial value" in the figure, the direction which gives the local optimal value of S/N ratio or sensitivity and the direction of the global optimal value (the point of "Optimal value" in the figure) are different. Therefore it is found that the true optimal value is not obtained by just a single analysis and it is necessary to perform the analysis repeatedly.

Therefore, as the means to solve this problem, an iterative application of Taguchi Method has been tried. In this application, the standard SN ratio method is applied repeatedly, while the optimal value of control factors determined in each analysis is set to the new middle level value for the next analysis. Figure 10.5 shows the concept of this iterative application. The optimal value of each control factor is determined from the main effect plot

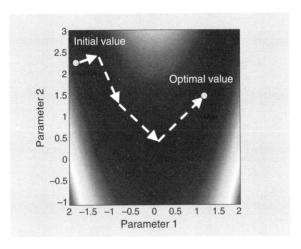

Figure 10.4   Example of strong interaction between control factors

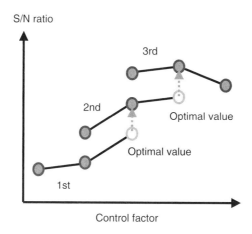

Figure 10.5 Concept of iterative application

obtained in the first analysis. In the second analysis, the optimal value is set to the new middle level. Then, the newly obtained optimal value is set to the third middle level and the same procedure is repeated. By this iteration, the optimal value of the control factor obtained in each analysis is expected to gradually approach the real optimal value. Though the vertical axis is SN ratio in Figure 10.5, the actual selection of the optimal value is done by comparing the main effect plots of SN ratio, $\beta 1$ and $\beta 2$.

The results after 10 iterations are shown in Figure 10.6, Figure 10.7, Figure 10.8, and Figure 10.9. Figure 10.6 shows a vibration spectrum before and after the optimization. It is found that the two main peaks are both 9 dB

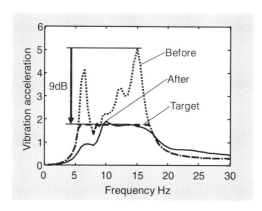

Figure 10.6 Vibration spectrum before and after iterative application

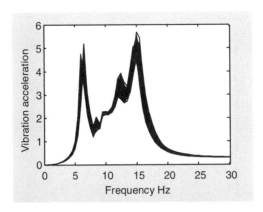

Figure 10.7   Variation of vibration spectrum before iterative application

reduced at the same time and that the design target is almost fully achieved. In the problem to reduce more than two vibration peaks like this, when one peak is decreased the other is often increased. But in this chapter, an excellent performance property is obtained without yielding any such phenomenon.

Figure 10.7 and Figure 10.8 show the dispersion of the vibration spectrum before and after iterative application. The curve group in each figure is the vibration spectrum varied by the noise factor. Before the optimization, the dispersion of the curve group is large and the influence of the noise factor to the vibration looks strong. However, it has become smaller after optimization and the influence of the noise factor is considered to be weakened. The S/N ratio has been improved by 13 db. Using the quality loss function, this means

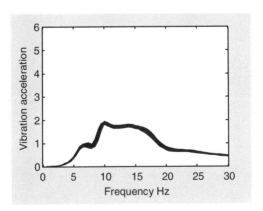

Figure 10.8   Variation of vibration spectrum after iterative application

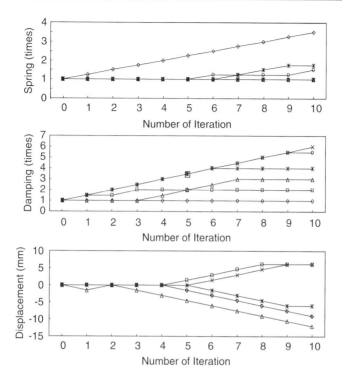

Figure 10.9   Change of control factors during iterative application

that defective incidence is reduced to $1/20$ if the system is defined to be defective when the magnitude of vibration exceeds twice the original value.

Figure 10.9 shows the change of each control factor due to iteration. It is found that many factors have unmonotonous change. For example, some of them do not change initially but start to change after some iteration, and on the contrary, another group does change first and stops it later. This is considered to be the effect of a large interaction between the control factors. The finally obtained value of the control factors are 0.25–1.5 times the original middle level value for the spring constant, 0.5 to 3 times for the damping coefficient, and 15–60 mm apart from the initial position.

10.6   Influence of Interval of Factor Level

Although in the previous analysis the level interval of the control factors is set to 5 times the maximum error, in order to examine the influence of the level

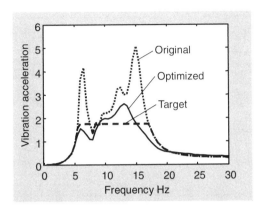

Figure 10.10   Change of vibration spectrum X3

interval to the convergence of the solution, additional analysis has been conducted with the following conditions:

- X3: level interval is 3 times the maximum error;
- X5: level interval is 5 times the maximum error;
- X10: level interval is 10 times the maximum error.

Figure 10.10 and Figure 10.11 show the vibration spectrum before and after the optimization for the case of X3 and X10 each. For X10, the target has been achieved in the whole frequency area, while for X3 the target has not been

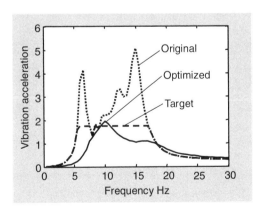

Figure 10.11   Change of vibration spectrum X10

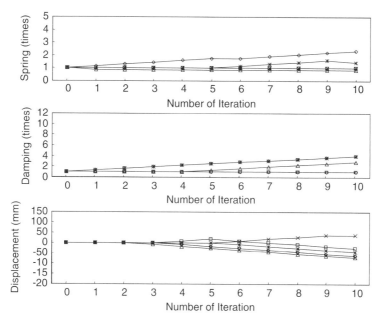

Figure 10.12   Change of control factors for X3

achieved in a part of the frequency area. It can be said that the convergence of the solution becomes slow when an excessively small interval is set like X3 and that the optimal solution may not be found due to the limitation of the computing time. Though the influences of the level interval to the final value and the convergence of S/N ratio were also examined, they were found to be comparatively small.

Figure 10.12 and Figure 10.13 show the change of the control factor for the case of X3 and X10 each. For X10, of which the level interval is large, the final value of some control factors becomes greater than double of that for X5, and those control factors looks to tend to diverge. From this fact, it is considered that if an excessively large interval is set, the solution may diverge and there is a possibility that a proper design result is not obtained. Considering the result for X3, it can be said that the level interval of the control factor should not be excessively small or excessively large and that it must be set to the moderate value.

In the Taguchi Method, in order to verify the optimal design result, the agreement of the improved quantity (gain) of the S/N ratio and the sensitivities estimated by Taguchi Method and those measured from the actual implementation of the optimal design are considered to be important

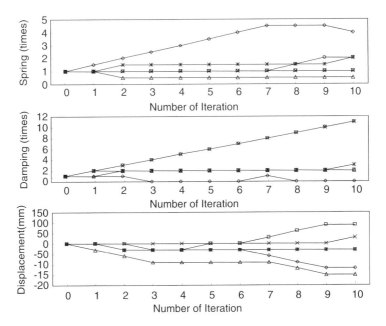

Figure 10.13 Change of control factors for X10

(reproducibility). Figure 10.14 shows the change of the estimated value and the measured value of the S/N ratio, $\beta 1$ and $\beta 2$ during the iteration. Here, the measured value means the S/N ratio and the sensitivities which are calculated through the simulation with the previously obtained optimal design result. In any case of X3, X5 and X10, the agreement of the estimated value and the measured value is fairly good and it can be considered that the proper optimal design result has been obtained. However, in the case of X10, since the solution tends to diverge, the estimated value and the measured value don't agree in some portions.

10.7 Calculation Program

In the optimal design based on an $L54 \times L54$ orthogonal array, about 3000 simulations are needed. If the analysis is repeated ten times, the number of simulations becomes 30 000 times. Therefore, to perform this analysis efficiently, the automatic calculation program is developed.

In this program, the standard S/N ratio Taguchi Method program written in the Excel macro invokes the engine vibration simulation program also

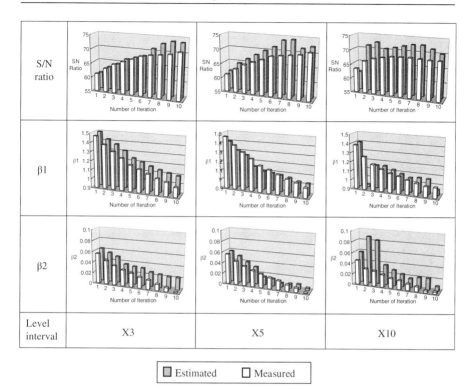

S/N ratio			
β1			
β2			
Level interval	X3	X5	X10

□ Estimated   □ Measured

Figure 10.14   Change of estimated and measured values of S/N ratio, β_1 and β_2

written in the Excel macro. The Taguchi Method program repeats determining the value of the design parameters using the orthogonal array, passing them to the simulation program, and retrieving the simulation result. Finally, the data for all simulation result is summarized, and the main effect plots of S/N ratio and sensitivity are made. The optimal value of each design parameters can be decided from those main effect plots.

By developing this program, the calculation efficiency of the optimal design method of this chapter has been improved greatly.

10.8 Conclusions

The standard S/N ratio Taguchi Method was applied to the optimal design of the three-point engine mounting system. The design of the mounting system with high robust performance has been derived by iteratively applying Taguchi Method with appropriately chosen factor level intervals. Namely,

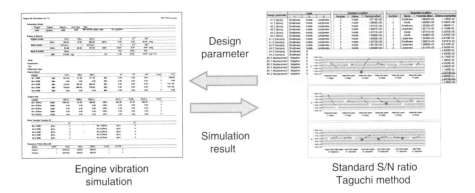

Engine vibration   Standard S/N ratio
simulation   Taguchi method

Figure 10.15 Calculation program

the excellent design has been obtained which can almost fully attenuate the main peaks of the low frequency vibration and at the same time reduce the defective incidence due to the manufacturing error to 1/20.

One of the future tasks is to establish the optimal design method to achieve a number of performance targets simultaneously. The targets will include also those for other than low frequency engine vibration.

10.8.1 Acknowledgments

The authors would like to thank Prof. Hiroshi Yano of Tokyo Denki University, the counsellor of Japanese Standards Association, for his guidance on this subject. The authors also appreciate the valuable advice of the Research Society on Manufacturing Functionality Evaluation in Hiroshima area.

10.1 References

1. Taguchi, Genichi (2001) Objective function and tuning. *Standardization and Quality Control*, **54** (8): 59–66 (in Japanese).
2. Taguchi, Genichi (2002) The measurement of the sound, S/N ratio and tuning. *Standardization and Quality Control*, **55** (10): 85–92 (in Japanese).
3. Taguchi, Genichi (2002) Basic function of sound and S/N ratio. *Standardization and Quality Control*, **55** (11): 81–90 (in Japanese).

This case study is contributed by Hiroshi Uchida, Hiromi Miwade, Kuniaki Nagao, Takahiko Tanaka, Naoto Takata, from Mazda Motor Corporation, Japan.

11

Optimization of a Front-Wheel-Drive Transmission for Improved Efficiency and Robustness

Chrysler Group, LLC, USA and ASI Consulting Group, LLC, USA

11.1 Executive Summary

This chapter illustrates a Design for Six Sigma project that was to optimize a front-wheel-drive automatic transmission for improved robustness and efficiency. In order to create improvements in a large complex system, it is key to take a large scope, to include as much of the system as possible. Five-phases of DFSS are Identify-Define-Develop-Optimize-Verify. This study considers the Identify-Optimize-Verify (IOV) phases. It was necessary to find a solution of the best transmission component setting that would not depend on noises

Robust Optimization: World's Best Practices for Developing Winning Vehicles,
First Edition. Subir Chowdhury and Shin Taguchi.
© 2016 Subir Chowdhury, Shin Taguchi, and ASI Consulting Group, LLC.
Published 2016 by John Wiley & Sons, Ltd.

such as oil temperature, or system pressures. The gear ratio was treated as an indicative factor in order to consider all driving conditions. To maximize the number of control factors evaluated utilizing hardware testing, Youden Square and Combination Design were used to modify the L18 orthogonal array. A dynamic ideal function was utilized focusing on power transformation, assessing over 98 000 design iterations against a wide spectrum of usage conditions in just four weeks of testing.

11.2 Introduction

A DFSS study was conducted to identify component design changes that would increase efficiency on a six-speed, front-wheel-drive automatic transmission. The transmission, the Chrysler 62TE, was chosen for the study based on its high production volume and the applications in which it is utilized. Initial studies have shown that by targeting specific components, significant contributions can be made to efficiency improvements. Since the transmission's design has already been established and in production for a number of years this DFSS project utilized the IOV approach to optimize its function of power transformation.

For this project Chrysler wanted to take into account the full spectrum of usage of this transmission to understand the contributions of each component modification in all driving conditions. The transmission efficiency changes significantly depending on driving conditions such as oil temperature, gear ratio, speed, load, and oil pressure. This study allowed Chrysler to identify any factors that could compromise efficiency at any driving condition. Chrysler did not want to improve efficiency for one condition while degrading another. Since different components are utilized for different gear ratios, variations in efficiency are expected and cannot be ignored. Therefore gear ratio was treated as an indicative factor.

The goal of this study was to optimize as many of the Front Wheel Drive (FWD) transmission components as possible to increase robustness and efficiency of the transmission. Youden Square and Combination Design used to modify the orthogonal array to increase the number of control factors while maintaining the lowest number of build combinations for hardware testing. While efficiency is important to the goal of this project, efficiency alone is insufficient for obtaining consistent improvements. Robustness, or consistency of efficiency under various driving conditions, is critical to achieving a better product. Efficiency and robustness must still be achieved while considering the cost impact.

11.3 Experimental

11.3.1 Ideal Function and Measurement

A dynamic ideal function was identified for this project based on the transmission's intended energy transformation. The transmission converts power from the input side (Torque x Speed) to power on the output side. It is important to look at this system dynamically because of the wide range of power input that a given transmission will see. Since the signal is part of the customer space, it is beneficial to expand the robustness to the maximum range of customer usage. Because input power consists of two elements, both speed and torque, that can be set independently, this ideal function has a double signal. It can be represented in Figure 11.1.

This energy transformation occurs for six different gears, four of which are considered in this study. While it is desirable to have robustness within each gear, differences between gears are expected due to the physics of the system. This lead the team to use gear ratio as an indicative factor, as will be fully discussed later. One Signal to Noise ratio is desired since the goal of the project is to have one optimal combination for all gears.

Figure 11.2 shows the ideal function with the indicative factor concept.

For this study we were able to utilize a dynamometer that could place a range of loads on the transmission at various speeds and conditions. Torque meters and speed sensors were located at each of the three dynamometers used. The layout of the test stand is represented in Figure 11.3.

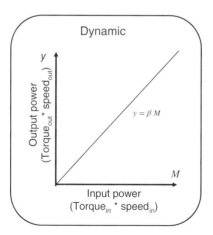

Figure 11.1   Ideal function

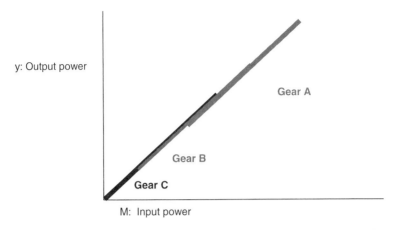

Figure 11.2   Ideal function displaying gear ratio as an indicative factor

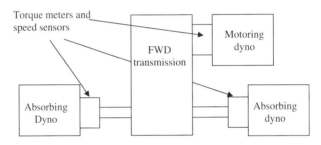

Figure 11.3   Test stand layout

An external water to oil cooler with a regulator valve was used to maintain the oil temperature at a constant value for all conditions. The pressure drop across the cooler represented a production level system. In order to focus our measurements on the transmission assembly alone, the torque converter clutch was engaged for all conditions.

11.4 Signal Strategy

The signal for this project was input power – the product of input speed and input torque. Inefficiencies are created when a clutch pack is in open condition. In order to capture all clutch packs in an open condition, a minimum of four gear ratios needed to be evaluated. Due to the configuration of the

Table 11.1   Signal strategy

Gear	Low		Medium		High	
	Input speed (RPM)	Output torque (Nm)	Input speed (RPM)	Output torque (Nm)	Input speed (RPM)	Output torque (Nm)
3	1600	503	2200	1006	2600	1711
4	1600	320	1760	634	2100	896
5	1500	234	1650	441	1900	750
6	1300	152	1450	304	1600	456

dynamometer, input speed and output torque were set, and input torque and output speed were measured to calculate the transmission efficiency. This is a unique aspect of the project, in which both the signal and response have measured and set elements. Though some of each were set and measured, the input power elements and output power elements were combined to be true to the energy transformation of the system.

A low input speed/low output torque, a medium input speed/medium output torque, and a high input speed/high output torque were used for the transmission's top four gear ratios. The remaining two gear ratios were not evaluated in this study because the components contributing to those gears also contributed to the top four. Using data collected from EPA City and Highway tests, the values for each of the three levels were determined for each gear. Table 11.1 shows the signal strategy.

11.5 Noise Strategy

For this study a compounded noise strategy was used. Historical test data has proven that transmission oil pressure, also considered line pressure, and oil temperature have a significant impact on the efficiency of the transmission. Since line pressure is required for the function of transferring power, it cannot be controlled to improve the efficiency. Oil temperature on the other hand can be controlled, but it is too expensive and out of the scope of this project. Higher line pressure and lower oil temperature result in a lower transmission efficiency value. Therefore those two factors were combined to create N1. Lower line pressure and higher oil temperature result in a higher transmission efficiency value, combined to create N2. Table 11.2 lists the values used in the compounded noise strategy for this project.

Table 11.2 Compounded noise strategy

Factor	N1 lower response	N2 higher response
Oil temperature	100 °F	180 °F
Line pressure	95 psi	70 psi

11.6 Control Factor Selection

All internal components of the 62TE transmission were in scope and available to be used as a control factor in this DFSS project. Figure 11.4 shows the common distribution of spin losses in an automatic transmission. Taking this into account, the project focused on clutch disks, separator plates, the oil pump and lubrication flow changes.

It was important to the team to be able to analyze the effects of each clutch pack. Thus, the seven sets of clutch disks and six sets of separator plates were identified as control factors. Another control factor chosen was a modified oil pump, since a typical pump contributes approximately half of the spin losses. Two additional control factors were selected to lower the fluid level in areas

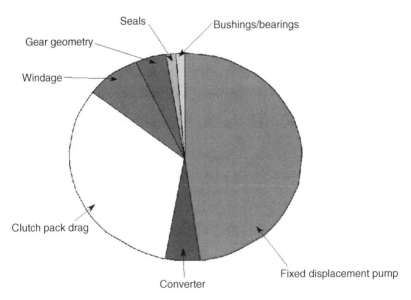

Figure 11.4 Distribution of spin losses in automatic transmission [1]

Table 11.3 Control factors and levels

	Control factor	Level 1	Level 2	Level 3
A	Oil level modification 1	Production	Prototype	
B	Oil level modification 2	Production	Prototype	
C	Clutch disk 1	Production	Prototype	
D	Clutch disk 2	Production	Prototype	
E	Clutch disk 3	Production	Prototype	
F	Separator Plate 1	Production	Prototype	
G	Clutch disk 4	Production	Prototype	
H	Separator Plate 2	Production	Prototype	
I	Clutch disk 5	Production	Prototype	
J	Separator Plate 3	Production	Prototype	
K	Clutch disk 6	Production	Prototype	
L	Separator Plate 4	Production	Prototype	
M	Clutch disk 7	Production	Prototype	
N	Separator Plate 5	Production	Prototype	
O	Pump modification	Production	Prototype	
P	Separator Plate 6	Production	Prototype 1	Prototype 2

with rotating components, thus reducing the churning losses in the transmission. The number of control factors selected for this project totaled 16. Of the control factors 15 had one prototype level in addition to the production comparison, and one control factor had two prototype levels. Table 11.3 lists the control factors and the levels used.

11.7 Orthogonal Array Selection

An orthogonal array was used to explore the design space of control factors. Each combination of control factors were tested against the noise strategy under each of the Torque x RPM configurations. The standard L18, shown in Table 11.4, allows for seven control factors with three levels and one control factor with two levels.

For this study, there were 16 control factors identified, 15 with two levels and 1 with three levels. Since this study was conducted with hardware, the most efficient array to use was a modified L18 using Youden Square and

Table 11.4 Standard L18

Run	A	B	C	D	E	F	G	H
1	1	1	1	1	1	1	1	1
2	1	1	2	2	2	2	2	2
3	1	1	3	3	3	3	3	3
4	1	2	1	1	2	2	3	3
5	1	2	2	2	3	3	1	1
6	1	2	3	3	1	1	2	2
7	1	3	1	2	1	3	2	3
8	1	3	2	3	2	1	3	1
9	1	3	3	1	3	2	1	2
10	2	1	1	3	3	2	2	1
11	2	1	2	1	1	3	3	2
12	2	1	3	2	2	1	1	3
13	2	2	1	2	3	1	3	2
14	2	2	2	3	1	2	1	3
15	2	2	3	1	2	3	2	1
16	2	3	1	3	2	3	1	2
17	2	3	2	1	3	1	2	3
18	2	3	3	2	1	2	3	1

Combination Design. Youden Square is a means of adding an additional column, and therefore control factor, at three levels [2]. It uses the degrees of freedom on the column 1 and 2 interaction of the standard L18 to help accomplish this, maintaining the overall degrees of freedom at 17. Combination Design is a means of rearranging the degrees of freedom of an orthogonal array to allow three-level columns to hold two separate two-level factors [2]. The orthogonality of the two new columns is compromised, but these columns remain orthogonal to the other columns of the array. In the technique, 3 of the 4 possible combinations of 2 factors are assigned to the original levels 1, 2, 3:

$1 = A1B1$
$2 = A1B2$
$3 = A2B1$

Table 11.5 Breakdown of array modification

Run	Original column	New column 1	New column 2
1	1	1	1
2	2	1	2
3	3	2	1
4	3	2	1
5	1	1	1
6	2	1	2
7	3	2	1
8	1	1	1
9	2	1	2
10	1	1	1
11	2	1	2
12	3	2	1
13	2	1	2
14	3	2	1
15	1	1	1
16	2	1	2
17	3	2	1
18	1	1	1

Table 11.5 breaks down the transformation of one 3 level column into two 2 level columns.

Some orthogonality is sacrificed when modifying the array. However, since a strong energy related ideal function was used, this sacrifice was warranted to study many factors. Using M: Power in and y: Power out is literally energy over time. Since energy is additive, the chances of strong interactions in the form of AxBxNoise are small.

The complete modified L18, shown in Table 11.6, allows for 15 control factors with 2 levels and 1 control factor with 3 levels, as desired.

The total experimental setup, including both control and noise factors, is shown in Table 11.7.

Table 11.6 Modified L18 ($3^1 \times 2^{15}$)

Run	A	B	C	D	E	F	G	H	I	J	K	L	M	N	O	P
1	1	1	1	1	1	1	1	1	1	1	1	1	1	1	1	1
2	1	1	1	1	2	1	2	1	2	1	2	1	2	1	2	1
3	1	1	1	2	1	2	1	2	1	2	1	2	1	2	1	1
4	1	1	2	1	1	1	1	1	2	1	2	2	1	2	1	2
5	1	1	2	1	2	1	2	2	1	2	1	1	1	1	1	2
6	1	1	2	2	1	2	1	1	1	1	1	1	2	1	2	2
7	1	2	1	1	1	1	2	1	1	2	1	1	2	2	1	3
8	1	2	1	1	2	2	1	1	2	1	1	2	1	1	1	3
9	1	2	1	2	1	1	1	2	1	1	2	1	1	1	2	3
10	2	1	1	1	1	2	1	2	1	1	2	1	2	1	1	2
11	2	1	1	1	2	1	1	1	1	2	1	2	1	1	2	2
12	2	1	1	2	1	1	2	1	2	1	1	1	1	2	1	2
13	2	1	2	1	1	1	2	2	1	1	1	2	1	1	2	3
14	2	1	2	1	2	2	1	1	1	1	2	1	1	2	1	3
15	2	1	2	2	1	1	1	1	2	2	1	1	2	1	1	3
16	2	2	1	1	1	2	1	1	2	2	1	1	1	1	2	1
17	2	2	1	1	2	1	1	2	1	1	1	1	2	2	1	1
18	2	2	1	2	1	1	2	1	1	1	2	2	1	1	1	1

11.8 Results and Discussion

11.8.1 S/N Calculations

Different transmission build combinations were evaluated to calculate the signal-to-noise ratio (S/N) as a measure of robustness, and the efficiency of the given design combination. A dynamic S/N was used to measure robustness since the transmission needs to be optimized over a range of input speeds and torques. The dynamic-signal-to-noise ratio assesses the engineered system's conversion of input signal into output response, as the levels of customer controlled noise factors vary in a systematic manner [3]. For the transmission, it is desirable to develop one S/N which quantifies the relative robustness of the given control factor combination across the desired signal range against both noise factor settings for all gears since, ultimately, one

Table 11.7 Experimental setup

	a	b	c	d	e	f	g	h	i	j	k	l	m	n	o	p
1	1	1	1	1	1	1	1	1	1	1	1	1	1	1	1	1
2	1	1	1	2	1	1	1	2	1	2	1	2	1	1	1	1
3	1	1	2	1	2	1	2	1	2	1	2	1	2	1	1	1
4	1	2	1	1	2	1	2	2	1	1	2	1	1	2	1	2
5	1	2	1	2	1	2	1	1	2	2	1	1	2	1	1	2
6	1	2	2	1	1	2	2	1	1	2	1	2	1	2	1	3
7	1	1	2	2	2	2	1	2	2	1	2	2	1	1	2	3
8	1	2	1	2	2	1	1	1	1	2	2	1	2	2	1	3
9	1	2	2	1	1	2	1	2	2	1	1	1	1	1	2	2
10	2	1	2	1	1	1	2	2	2	2	2	2	1	1	2	2
11	2	1	1	2	2	2	2	1	1	1	1	1	2	1	2	2
12	2	2	2	1	2	1	1	1	2	2	1	2	1	1	1	3
13	2	1	1	2	1	2	2	2	1	1	2	1	1	2	1	3
14	2	2	2	1	1	1	1	2	2	2	1	1	2	1	1	1
15	2	1	2	2	2	2	1	1	1	1	2	2	1	2	1	1
16	2	2	1	1	2	2	2	1	2	2	1	1	1	1	2	1
17	2	1	2	1	1	1	2	1	1	1	2	2	1	1	1	1
18	2	2	1	2	1	1	1	1	2	2	1	1	2	1	1	1

3rd Gear							4th Gear							5th Gear							6th Gear							3rd Gear		4th Gear		5th Gear		6th Gear		Overall	
N1			N2				N1			N2				N1			N2				N1			N2				SN	B	SN	B	SN	B	SN	B	SN	B
L/L	M/M	H/H	L/L	M/M	H/H		L/L	M/M	H/H	L/L	M/M	H/H		L/L	M/M	H/H	L/L	M/M	H/H		L/L	M/M	H/H	L/L	M/M	H/H											

Table 11.8 Experimental set up for each build combination

		M1 Low speed low torque	M2 Medium speed medium torque	M3 High speed high torque
N1 Low temperature High line pressure	3rd Gear	y1	y2	y3
	4th Gear	y4	y5	y6
	5th Gear	Y7	Y8	Y9
	6th Gear	y10	y11	y12
N2 High temperature Low line pressure	3rd Gear	y13	y14	y15
	4th Gear	y16	y17	y18
	5th Gear	y19	y20	y21
	6th Gear	y22	y23	y24

design solution is preferred across all of these elements. Conceptually, for this transmission design, the S/N is:

$$S/N = 10\log\left(\frac{\beta^2}{V_{Noise}}\right), \tag{11.1}$$

where β = the overall efficiency and V_{Nosie} = variance due to the noise factors. Table 11.8 displays the experimental layout for each build combination.

The outer array analysis of variance decomposition for this project is as follows:

$$S_T(f_T = 24)\begin{cases} S_L(f_L = 8) & \begin{cases} S_\beta(f_\beta = 1) \\ S_{\beta\times N}(f_{\beta\times N} = 1) \\ S_{\beta\times G}(f_{\beta\times G} = 3) \\ S_{\beta\times N\times G}(f_{\beta\times N\times G} = 3) \end{cases} \\ S_e(f_e = 16) \end{cases} \tag{11.2}$$

Since the goal is to increase the overall efficiency of the entire system for all gears, S_β is what is to be maximized.

$$S_{\beta_{total}} = \frac{\left(\sum_{i=1}^{n} M_i \times y_i\right)^2}{r} \tag{11.3}$$

$$r = \sum_{i=1}^{n} M_i \tag{11.4}$$

Note that:

$$\beta^2 = \frac{1}{r}S_{\beta_{total}} \qquad (11.5)$$

The effect of the noise needs to be minimized and included in the denominator of the S/N. The denominator is represented by the variance due to noise, or the spread and nonlinearity in the ideal function that results when noise factors steal or misplace energy from the desired ideal energy transformation. This includes the effect of the compounded noise, N1 and N2, but not the changes that result from the gear ratio. Since different components within the transmission are used for the different gear ratios, the variation that results due to each gear is expected and acceptable. This effect needs to be ignored. Therefore, gear is an indicative factor. An indicative factor is defined as a factor with levels of customer usage, but the *variation* that results from this factor on the ideal function is ignored or removed [4]. This effect needs to be extracted from the data. So of the above elements, $S_{\beta\times N}$, $S_{\beta\times N\times G}$, S_e all need to be minimized, while $S_{\beta\times G}$ should be ignored.

When evaluating gear ratio as an indicative factor, traditionally the variance due to noise would be calculated using linear equations as:

$$L1 = M1y1 + M2y2 + M3y3 \qquad (11.6)$$

$$L2 = M1y4 + M2y5 + M3y6 \qquad (11.7)$$

. . . and

$$S_{\beta\times N} = \left(\frac{L1 + L2 + L3 + L4}{4xr}\right)^2 + \left(\frac{L5 + L6 + L7 + L8}{4xr}\right)^2 - S_\beta \qquad (11.8)$$

$$S_{\beta\times G} = \left(\frac{L1 + L5}{2xr}\right)^2 + \left(\frac{L2 + L6}{2xr}\right)^2 + \left(\frac{L3 + L7}{2xr}\right)^2 + \left(\frac{L4 + L8}{2xr}\right)^2 - S_\beta \qquad (11.9)$$

with

$$V_{noise} = (S_T - S_\beta - S_{\beta\times G})/(n - f_\beta - f_{\beta\times G}) \qquad (11.10)$$

or

$$V_{noise} = V_{\beta\times N} + V_{\beta\times N\times G} + V_e \qquad (11.11)$$

Since the signal range is different for each gear and slightly different for each noise, linear equations can not be used with the measured data as is. Therefore, the best way to represent the SN is:

$$S/N = 10\log\left(\frac{\frac{1}{r}S_{\beta total}}{\frac{Ve_3 + Ve_4 + Ve_5 + Ve_6}{4}}\right) \tag{11.12}$$

where Ve_3, Ve_4, Ve_5, and Ve_6 (Ve_j) are the variances due to noise for the data for each individual gear (j). This is represented by:

$$Ve_j = \left(\frac{S_{T_j} - S_{\beta_j}}{n_j - 1}\right) \tag{11.13}$$

where

$$S_{T_j} = \sum_{k=1}^{n_j} y_k^2 \tag{11.14}$$

$$S_{\beta_j} = \frac{\sum\limits_{k=1}^{n_j}\left(M_k \times y_k\right)^2}{\sum\limits_{k=1}^{n_j} M_k} \tag{11.15}$$

For each combination of control factors, the S/N and β were calculated. Overall efficiency is given by:

$$\beta_{overall} = \frac{\left(\sum\limits_{i=1}^{n} M_i \times y_i\right)}{r} \tag{11.16}$$

where r is defined in Equation (11.4).

A sample calculation for one combination of the L18 is shown in Table 11.9.

11.8.2 Graphs of Runs

A graph of this run can be seen in Figure 11.5. This graph shows the power in and power out measurements for all gears, signal levels, and noise factors that were measured for one of the build combinations for this project.

Table 11.9 Sample calculation for a combination of the L18

Overall		3rd gear	
R	1.47 e + 12	r 3	6.11 e + 11
Sβ total	1.2 e + 12	Sβ3	5.01 e + 11
Ve average	34430079	Ve 3	14775174
S/N	−76.3	ST 3	5.01 e + 11

Once the S/N and β were calculated for each run, the average effects of each control factor and level were calculated as they pertained to both S/N and β. This was done per the orthogonal array, where for a given factor and level, S/N values from the runs with those respective factor level settings were averaged. This same process was followed for β.

11.8.3 Response Plots

Figure 11.6 shows the S/N values calculated for each control factor and level. The values circled are the control factors to focus on for improvements in robustness. The average S/N value plotted represents the average S/N value for all 18 runs conducted in the experiment.

Figure 11.7 shows the β results for each of the control factors and their levels. The averaged values for all of the runs were zeroed out, and this plot shows the absolute efficiency improvements or degradation. The values circled highlight control factors to be pursued for efficiency improvements.

11.8.4 Confirmation Run

After completing the 18 runs and calculating the S/N and β for all 16 control factors and each of their levels, the Taguchi two step was completed and an optimal build combination was determined. The combination used for the optimal build is listed in Table 11.10. The predicted S/N improvement for that specific combination was 1.89 dB and the predicted β improvement was 0.88%. A transmission was built to the optimal build combination and run on the test stand at the same conditions as the original 18 builds. The test resulted in a 1.27 dB gain, which is equivalent to a 13.6% reduction in variation and a 0.60% efficiency improvement. Since the optimal build resulted in 70% of the predicted gain the results are believed to be repeatable. Any effects based on additional noise factors not included in this study, or effects based on measurement error are minimal and insignificant.

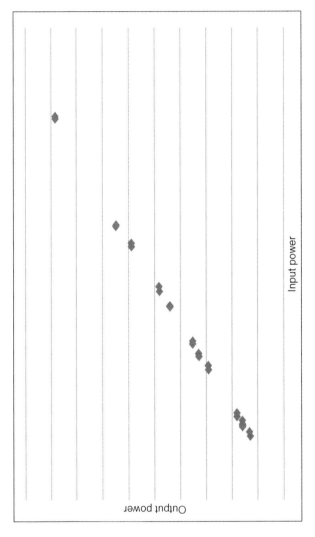

Figure 11.5 Data from a combination of the L18

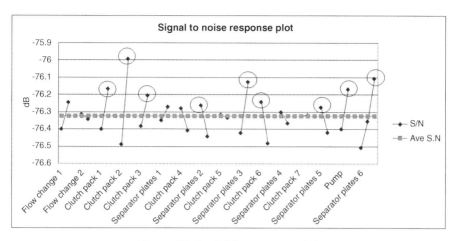

Figure 11.6   S/N response plot

11.8.5 Verification of Results

The 18 optimization evaluations and the confirmation test were performed in such a way as to test and understand the design iterations quickly and minimize the amount of data collected. To ensure these improvements will be seen in a larger scale and in the real world, the optimal design was evaluated under standardized test conditions. The Optimization Phase resulted in an average efficiency improvement of 0.6% under the conditions

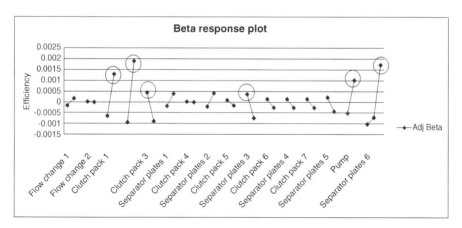

Figure 11.7   Beta response plot

Table 11.10 Optimal build combination

	Control factor	Optimal build
A	Fluid level modification 1	Level 2
B	Fluid level modification 2	Level 1
C	Clutch disk 1	Level 2
D	Clutch disk 2	Level 2
E	Clutch disk 3	Level 1
F	Separator Plate 1	Level 2
G	Clutch disk 4	Level 1
H	Separator Plate 2	Level 1
I	Clutch disk 5	Level 1
J	Separator Plate 3	Level 2
K	Clutch disk 6	Level 1
L	Separator Plate 4	Level 1
M	Clutch disk 7	Level 1
N	Separator Plate 5	Level 1
O	Pump modification	Level 2
P	Separator Plate 6	Level 3

identified in Table 11.1 and 11.2. The Verification Phase utilized conditions identified in Table 11.11 for each gear at 200 °F and production line pressure values. These conditions represent a larger range of driving conditions to fully understand the benefit. The data for the Verification Phase was collected utilizing the same loaded dynamometer test cell that was used for the initial phases of the project. The average efficiency improvements measured for all gears, loads, and speeds in this phase resulted in a 2.19% gain over the baseline build. Because the driving conditions for the Optimize and Verification Phases are different, it was to be expected that the efficiency gains vary as well. It is important to identify that both phases show a significant gain and are directionally correct.

In order to understand how these efficiency improvements relate to overall fuel economy improvements, a simulation was completed utilizing software developed by Chrysler engineers.

Table 11.11 Verification test conditions

Engine speed (RPM)	Input torque (Nm)																
	13.55	20.34	27.12	40.67	54.23	67.79	101.69	135.58	169.48	203.37	237.27	271.16	305.06
500													
750													
1000													
1250													
1500													
1750													
2000													
2250													
2500													
2750													
3000													

11.9 Conclusion

This study showed how a front-wheel-drive-automatic transmission can be optimized utilizing the IOV phases of DFSS. A dynamic ideal function measuring the power transformation was applied. This project identified an efficiency improvement of 0.60% and a variation reduction of 13.6%. By focusing on key enablers and utilizing Youden Square and Combination Design to modify an L18 orthogonal array, hardware testing was completed in only four weeks. Modifying the orthogonal array allowed this project to evaluate 16 control factors, compared to eight with a standard L18, in only 18 runs. This is a huge in testing efficiency. In DFSS, it is critical that as many control factors as possible be studied, especially for transmissions. Every opportunity for Robustness and Efficiency, and cost savings is needed to gain competitive advantage. Modifying the array sacrificed some orthogonality, but with a strong energy focused ideal function (energy being additive) it was acceptable.

Also key to the efficiency of this project was the use of only three signal levels for each gear and noise combination. While many more combinations of Torque and RPM exist in actual application of the transmission, using only three of the combinations that *span the range of usage* is sufficient for understanding both Robustness and overall efficiency. This minimalist technique savings both time and money and drove ultimate improvement.

This project also had to deal with the fact that both the signal and the response was partially measured, partially set. Using the average of the noise variances for each of the gears allowed this analysis to be easily done. It is important to understand the what the decomposition of the outer array truly represents to be able to develop this type of S/N.

Efficiency variation is expected and accepted from gear to gear. Because of this the gear ratio was treated as an indicative factor. Although each gear ratio produced different results, only one S/N value was calculated since only one optimal build combination can be used for the transmission for all driving conditions. The optimal build combination predicted from the 18 test runs resulted in a 2.19% efficiency improvement over the production baseline build. Since the driving conditions varied from the testing done in the Optimal Phase to the Verification Phase, the difference in efficiency improvements are to be accepted. Both phases showed significant gains.

11.9.1 Acknowledgments

The authors would like to acknowledge our families for supporting us in writing this chapter. We would also like to thank Dan Black, Mircea Gradu, Genichi Taguchi, Shin Taguchi, and Subir Chowdhury for their wisdom, mentorship, and support as they continue to allow and encourage us to grow. We would also like to thank the coworkers at Chrysler's automatic transmission lab for their assistance with all of the testing.

11.10 References

1. Kluger, M. and Greenbaum, J.J. (1993) Automatic transmission efficiency characteristics and gearbox torque loss data regression techniques. *Society of Automotive Engineers, Paper No. 930907.*
2. DFSS Master Blackbelt Expert Course Pack, *Modifying Orthogonal Arrays*, ASI Consulting Group.
3. Jensen, C., Quinlan, J. and Feiler, B. (2008) Robust engineering and DFSS: how to maximize user delight and function and minimize cost, *Society of Automotive Engineers, Paper No. 2008-01-0361.*
4. Wu, Y. and Wu, A. (2000) *Taguchi Methods for Robust Design.* New York: ASME Press.

This case study is contributed by Lauren L. Thompson of Chrysler Group, LLC and Craig D. Jensen of ASI Consulting Group, LLC.

12

Fuel Delivery System Robustness

Ford Motor Company, USA

12.1 Executive Summary

Excessive fuel returned from the combustion area back to the gas tank is no longer allowed because of air pollution. This phenomenon is called hot fuel handling. Because an excess of fuel is not allowed, there is a strong need to develop a fuel pump system which delivers accurate fuel flow. Fuel flow is controlled by electronics. Electronic control is the future of the engineered system. For any manufacturing company, it is important to establish an effective product development strategy to develop a robust electronic control system. The project team had three years to develop the new system. Since they progressed much even after one and half years, the project was started.

Instead of evaluating the fuel delivery system by testing it to find problems, a robust design approach was used. This approach evaluates the design by measuring the variability of the energy transformation of the system. It looks for design that has the least variability of energy transformation regardless of a customer's usage.

Robust Optimization: World's Best Practices for Developing Winning Vehicles,
First Edition. Subir Chowdhury and Shin Taguchi.
© 2016 Subir Chowdhury, Shin Taguchi, and ASI Consulting Group, LLC.
Published 2016 by John Wiley & Sons, Ltd.

Instead of repeating the design-build-test cycle several times until the design is good enough, the robust design approach systematically optimizes the design by studying several design parameters simultaneously. The team spent most of the energy to discuss what should be measured and how it should be tested. It took several four-hour meetings to plan the testing.

Five design parameters were tested under two noise conditions which were generated by varying fuel type, fuel temperature and tank pressure, and by indicating variability in customer usage conditions.

12.2 Introduction

Today's vehicles require the fuel system performance to be consistent and predictable over the operating range regardless of environmental conditions. This study provides a method by which Ford Motor Company identified the components that yield a robust fuel delivery system.

Inconsistent liquid fuel at the rail may cause customer dissatisfaction manifested in the following vehicle symptoms:

- difficult restart after engine-off soak;
- rough or rolling engine idles;
- engine stumbling while cruising, accelerating and decelerating.

Insufficient liquid fuel at the fuel injector is a root cause for these symptoms experienced by customers in the field. Higher temperatures in the engine, underbody, and fuel tank as well as high volatility fuel lead to fuel vaporization which may cause insufficient fuel delivery to the injectors. The higher temperatures and fuel characteristics cannot be controlled or specified by engineering so they are considered noise factors. A robust system or subsystem must perform consistently regardless of the noise factors to which it is exposed.

This study examines the fuel delivery subsystem to specify the fuel delivery components that contribute to a robust fuel system.

12.2.1 Fuel System Overview

Ford Motor Company continually strives for improved quality and new technologies to increase customer satisfaction. A new fuel system being developed targets increased fuel system robustness in the varied environmental conditions to which the vehicle is exposed. Brief overviews of the existing conventional fuel system and the new fuel system follow.

12.2.2 Conventional Fuel System

The conventional system applies battery voltage to the pump and regulates the fuel rail differential pressure via a mechanical regulator. The fuel not consumed by the engine is returned to the tank. The typical operating range for the fuel pump is 12–14 volts and 200–300 kPa.

12.2.3 New Fuel System

Ford Motor Company is developing a fuel system which regulates the pump speed and pressure. The speed and pressure range of the fuel pump is greater with this new fuel system than with the conventional system used in all Ford cars and light trucks. The typical operating range for the fuel pump is 6–13 volts and 200–400 kPa. The fuel pressure in the new system is maintained through an electronic controller via closed-loop feedback. The controller contains a table of the fuel pump performance with flow as a function of pressure and voltage. When the actual pressure differs from that desired, the controller adjusts the pump voltage to compensate for the pressure difference.

The pump flow as a function of voltage and pressure must be accurate throughout the operating range and under all conditions to minimize the error for which the electronic controller must compensate.

12.3 Experiment Description

12.3.1 Test Method

Parameter design using Taguchi's design of experiment method uses a system that aspires to an ideal function of performance. The ideal function is the measured response to a system input signal. All external and environmental factors are considered noise. Comparing system performance when subjected to favorable and unfavorable noises allows assessment of system robustness. A robust system will minimize the performance differences when subjected to the various noises.

12.3.2 Ideal Function

The ideal function for this study is based on physics. It is derived from fuel pump efficiency. Pump voltage was varied to change the power supplied to the fuel pump. Fuel pump current was observed and multiplied by the pump voltage to represent the system input signal. Flow rate was observed as the output response at four back pressure levels and five voltage levels.

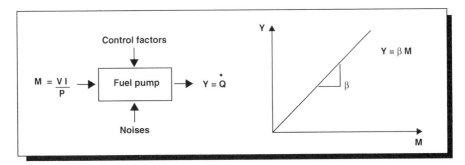

Figure 12.1   P-diagram and ideal function

Figure 12.1 contains a schematic of the engineered system and the ideal function.

$$\text{Pump efficiency: Efficiency} = \frac{\dot{Q} \times P}{V \times I} \tag{12.1}$$

$$\text{System response: } Y = \dot{Q} \tag{12.2}$$

$$\text{System input signal: } M = \frac{V \times I}{P} \tag{12.3}$$

$$\text{System ideal function: } Y = \beta M \tag{12.4}$$

where

β = efficiency
$\dot{Q}$ = fuel pump flow rate
V = fuel system back pressure
I = fuel pump current.

The levels of the input signals V and P are in Table 12.1 and Table 12.2. The fuel pump in the new fuel system is exposed to the ranges of voltages and pressures listed.

Table 12.1   Fuel pump voltage levels

	V_1	V_2	V_3	V_4	V_5
Pump voltage (volts)	6	8	10	12	14

Table 12.2 Fuel pump pressure levels

Test pressure (kPa)	P_1	P_2	P_3	P_4
	200	250	300	350

12.4 Noise Factors

As mentioned above, noise is an uncontrolled environmental condition. The noise factors and their compounded levels are listed in Table 12.3. The noise factors were selected to specifically address hot fuel handling issues. Noise factors level 1 are unfavorable to the fuel delivery system; they tend to lower the pump's flow. Noise factors level 2 are favorable: they tend to increase the pump's flow.

Fuel volatility (RVP values) represents the range of commercially available fuels in summer months. Higher RVP indicates increasing fuel volatility. Higher volatility fuels allow more vapor generation, which makes the fuel delivery system more susceptible to being unable to provide liquid fuel to the fuel injectors.

Fuel temperature values are the range of in-tank fuel temperatures expected during the summer months. Higher fuel temperatures have the same effect as higher volatility fuels; they increase the fuel vaporization.

The tank vapor pressure range is based on government regulations. Higher tank vapor pressures will have the tendency to increase fuel pump flow since there is an additional static fluid head at the fuel pump inlet.

Table 12.3 Noise factors and levels

Noise factors	Fuel type (Reid vapor pressure, RVP)	Fuel temperature (°C)	Tank vapor pressure (in. H_2O)
N1 (low pump flow)	High	High	0
N2 (high pump flow)	Low	Low	10

12.4.1 Control Factors

Unlike noise factors, engineering can influence control factors. The control factors chosen for this experiment are specified or recommended by fuel delivery engineering as shown in Table 12.4. Note that selection of control factors is limited as this study was done by an OEM. Idealy, more control factors should be taken from the design of the fuel pump itself.

Table 12.4   Control factors and levels

Control factors	Level 1	Level 2	Level 3
A. Fuel pump type	Turbine	Gerotor	N/A
B. Assembly type	Fuel delivery module (FDM)	Bracket	Bracket with jet pump
C. Mounting angle (from vertical)	0	45	80
D. Rated pump flow (Lph)	Low	Medium	High
E. Modulation frequency (kHz)	4	9.6	19.2

12.4.2 Fixed Factors

The following fuel system factors are fixed:

- fuel level in tank;
- fuel heating rate;
- fuel tank size;
- fuel filter sock and its orientation to fuel in the tank.

12.5 Experiment Test Results

Table 12.5 summarizes the results from L18.

Since our goal is to minimize variations when exposed to the different noise factor levels, we are primarily concerned with maximizing the S/N ratio. The effects of the control factors on S/N are shown in Figure 12.2 and summarized in Table 12.6. A sensitivity analysis (analysis on b) discussion follows.

12.6 Sensitivity (β) Analysis

Our customer is more concerned with predictable flow to the engine than with the overall efficiency of the fuel delivery system. For that reason, the sensitivity analysis is secondary in this study. The differences between the optimal system for maximizing S/N ratio and the system maximizing b are control factors B and D. Figure 12.3 and Table 12.7 show the control factor level results for sensitivity.

Table 12.5 Experiment design and results

Test	A	B	C	D	E	S/N	Beta
1	1	1	1	1	1	25.42	415
2	1	1	2	2	2	23.04	260
3	1	1	3	3	3	22.90	400
4	1	2	1	1	2	28.21	493
5	1	2	2	2	3	22.04	387
6	1	2	3	3	1	22.57	423
7	1	3	1	2	1	17.10	255
8	1	3	2	3	2	20.55	391
9	1	3	3	1	3	23.79	352
10	2	1	1	3	3	20.19	348
11	2	1	2	1	1	17.48	222
12	2	1	3	2	2	22.22	384
13	2	2	1	2	3	19.59	363
14	2	2	2	3	1	14.20	316
15	2	2	3	1	2	18.61	346
16	2	3	1	3	2	22.17	430
17	2	3	2	1	3	17.10	255
18	2	3	3	2	1	20.62	366

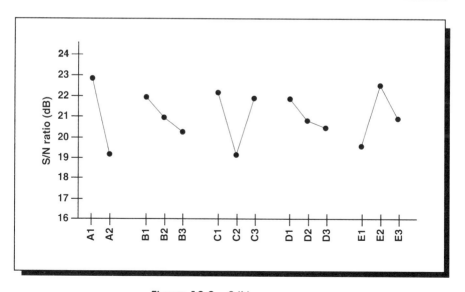

Figure 12.2 S/N responses

Table 12.6 Control factors

	Control factor A	Control factor B	Control factor C	Control factor D	Control factor E
Level 1	22.85	21.88	22.11	21.80	19.57
Level 2	19.15	20.90	19.07	20.77	22.50
Level 3		20.22	21.82	20.43	20.94

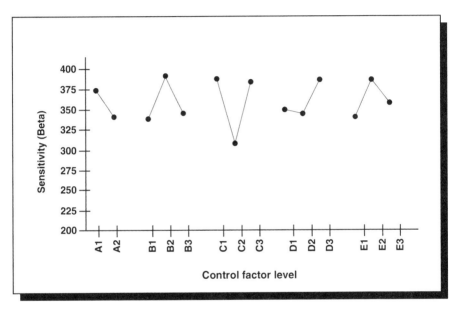

Figure 12.3 Sensitivity β analysis

Table 12.7 Control factor results

	Control factor A	Control factor B	Control factor C	Control factor D	Control factor E
Level 1	375	338	387	347	338
Level 2	341	391	305	343	384
Level 3		345	382	385	354

Our system response is the fuel flow to the engine. Consistent fuel flow to the engine is our customer's expectation from the fuel delivery system. Two of the fuel delivery assembly control factors divert fuel pump outlet fuel flow from the engine into the fuel tank. This will decrease the efficiency of the system. Since b is proportional to pump efficiency, b decreases with the systems that divert the flow to the tank. This is evident in control factor B. Level B2 is considerably higher than B1 or B3.

$$T = \text{Overall average of S/N} = 21.0.$$

The optimal fuel delivery system to maximize S/N is:

Pump type:	Turbine
Assembly type:	FDM
Mounting angle:	00 from vertical
Pump flow:	Low
Modulation frequency:	9.6 kHz

12.7 Confirmation Test Results

We performed confirmation and verification tests to validate the L18 experimental results. They are divided into two sections:

- Bench tests;
- Vehicle tests.

For each set of tests the initial design is compared to the optimal design as determined by the L18 orthogonal array.

The initial configuration fuel delivery system is:

Pump type:	Generator
Assembly type:	Bracket
Mounting angle:	800 from vertical
Pump flow:	Low
Modulation frequency:	19.2 kHz

12.7.1 Bench Test Confirmation

The two system configurations were tested using the same test procedure and method.

12.7.1.1 Initial Fuel Delivery System

The predicted S/N for the initial fuel delivery system is 20.6 dB as calculated in Equation (12.5).

$$\eta_{initial} = \overline{T} + (\overline{A_2} - \overline{T}) + (\overline{B_2} - \overline{T}) + (\overline{C_3} - \overline{T}) + (\overline{D_1} - \overline{T}) + (\overline{E_3} - \overline{T}) = 20.96 dB.$$

$$(12.5)$$

12.7.1.2 Optimal Fuel Delivery System

The predicted S/N for the most robust fuel delivery system is 27.1 dB as calculated in Equation (12.6).

$$\eta_{optimal} = \overline{T} + (\overline{A_1} - \overline{T}) + (\overline{B_1} - \overline{T}) + (\overline{C_1} - \overline{T}) + (\overline{D_2} - \overline{T}) + (\overline{E_2} - \overline{T}) = 27.1 dB.$$
(12.6)

Prediction and the result from confirmation are summarized in Table 12.8.

The bench confirmation tests yielded a 9.1 dB increase from the original fuel delivery system to the optimal one.

The difference between the actual and predicted values for S/N in the confirmation test indicates the possibility of some interactions and/or some unknown noise factor(s).

12.7.2 Vehicle Verification

Verification in the vehicle is the successful completion of a hot fuel handling test. The test outline is:

- Warm up/grade load/engine soak
- City drive
- Extended idle/engine soak

The fuel delivery system is exposed to a condition similar to noise level 2 in the beginning of the test. As the test progresses the condition in the fuel tank changes toward noise level 1. That may cause insufficient liquid fuel at the fuel rail. When there is insufficient liquid fuel, the symptoms of a nonrobust fuel delivery system may occur.

Table 12.8 Prediction and confirmation

	Prediction	Confirmation
Initial design	20.6 dB	14.9 dB
Optimal design	27.1 dB	24.0 dB
Gain	6.5 dB	9.1 dB

12.7.2.1 Initial Fuel Delivery System

The initial system did not pass the hot fuel handling test. During the grade load section the engine stalled due to insufficient fuel pressure. Figure 12.4 contains a plot of the actual fuel pressure compared to the requested fuel pressure. Note the increasing difference between the requested and actual pressure. The actual pressure fluctuations are due to vapor handling difficulty at the fuel pump.

12.7.2.2 Optimal Fuel Delivery System

The optimal fuel delivery system successfully passed the hot fuel handling test. Figure 12.5 is the plot of actual fuel pressure and requested fuel pressure during the grade load part of the test. The requested and actual pressure signals overlap with the optimal system indicating that the optimal system provides the pressure required.

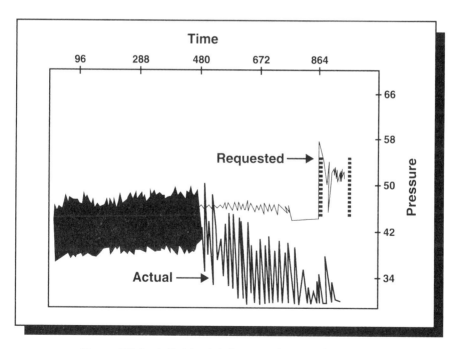

Figure 12.4   Initial fuel delivery system performance

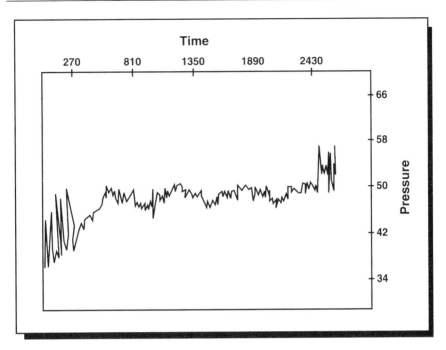

Figure 12.5   Optimal fuel delivery system performance

12.8 Conclusion

Parameter design for a robust delivery system indicates that the following parameters yield the optimal delivery system:

- Turbine fuel pump
- Fuel delivery module
- 00 from vertical orientation
- Lowest pump capacity
- 79.6 kHz pump modulation frequency.

The optimal design proved to have better performance than the initial design in both bench testing and vehicle evaluation.

The primary benefit associated with this approach to problem solving is increased customer satisfaction. Our internal Ford customers are more satisfied since they can continue vehicle development without concerns regarding the fuel delivery system. Our customers will also be more satisfied due to the decreased likelihood of hot fuel handling issues while they own a Ford vehicle. We estimate that 20% of our field returns will be avoided using the optimal fuel delivery system.

Other benefits are decreased development time and cost. An alternative approach used to address this issue consisted of multiple bench studies followed by vehicle studies. That approach resulted in the same conclusion; however, it took approximately six months longer due to learning in smaller incremental steps, and it cost an additional $25000.

This case study is contributed by J.S. Colunga, D. Lau, J.R. Otterman, B.C. Prodin, K.W. Turner, and J.J. King, of Ford Motor Company, USA.

13

Improving Coupling Factor in Vehicle Theft Deterrent Systems Using Design for Six Sigma (DFSS)

General Motors Corporation, USA

13.1 Executive Summary

PASS-Key III+ (PKIII+) is the second generation of transponder based immobilizer systems in General Motors' vehicles. It has enhancements over PASS-Key III (PKIII) in that it is General Motors first cryptographic immobilizer which means the signal from the key is both encrypted and randomized. This system has proven very effective in deterring theft.

The system's ability to communicate between the vehicle and transponder in the key is measured by the coupling factor.

Robust Optimization: World's Best Practices for Developing Winning Vehicles,
First Edition. Subir Chowdhury and Shin Taguchi.
© 2016 Subir Chowdhury, Shin Taguchi, and ASI Consulting Group, LLC.
Published 2016 by John Wiley & Sons, Ltd.

The focus of this DFSS project is to understand the physics and critical design parameters involved in achieving optimal coupling factor to improve the first time quality in future designs. Achieving this objective will lead to designs robust to variances in material and packaging design and result in less testing.

The process used in the past on these systems was the Design-Test-Fix approach. Using an orthogonal array to understand the factors critical to coupling factor was a new approach for this system.

An L18 orthogonal array was utilized to assess various factors and their associated levels.

13.2 Introduction

General Motors is the largest manufacturer of automotive vehicles in the world. As such, GM's products are a desirable target for car thieves particularly in North America where GM has the largest market share of all manufacturers. Circa 1990, GM began to provide higher theft rate vehicles with engine immobilizers.

People's vehicles are an important investment and extension of their personal lives. The security of their vehicles is an important customer requirement which manufacturers need to address. Therefore, vehicle security is important to both General Motors and its customers.

GM's initial foray into the vehicle theft deterrent arena came in the form of VATS (Vehicle Anti-Theft System) which was introduced on the Corvette in 1986. In this system, a resistor is built into the key and the system looks for the correct resistor value in the key. If the key has the correct resistor value and can turn the mechanical lock then the VATS module will send a code to the engine control module allowing the vehicle to start. By 1989, GM began rolling out this system in the form of PASS-Key I or PASS-Key II across its higher theft rate vehicle lines and theft rates in those vehicles declined. These systems have 15 different resistor values and the system will respond to an incorrect resistor value by locking out any further attempts to start the vehicle for 2–3 minutes at which point it will allow another attempt.

The introduction of transponder based vehicle theft deterrent systems in General Motors products began in 1997. The transponder based immobilizer system carries advantages of both of the previous generations of systems: (1) it has the advantage of the VATS/PASS-Key I/II systems in that the driver takes away a portion of the system when they walk away from the vehicle with the key; and (2) it has the advantages of the Passlock system in that any key blank can be made to work within any given vehicle with the correct mechanical cut and programming.

PASS-Key III+ (PKIII+) is the second generation of transponder based immobilizer systems. It has strengths over PKIII in that it is General Motors first cryptographic immobilizer which means the signal from the key is both encrypted and randomized. This system has proven very effective in deterring theft.

13.3 Objectives

The system's ability to communicate between the vehicle and transponder in the key is measured by the coupling factor.

The focus of this DFSS project is to understand the physics and critical design parameters involved in achieving optimal coupling factor to improve the first time quality in future designs. Achieving this objective will lead to designs robust to variances in material and packaging design and result in higher success in initial validation testing.

13.4 The Voice of the Customer

We used Quality Function Deployment (QFD) to translate the Voice of the Customer into product requirements (Figure 13.1). Through analysis, it was determined that Coupling Factor and Signal Quality would be the focus and be carried to the next level QFD house.

The second level of QFD brings us to a more actionable level. Here we see various factors impact on coupling factor and signal quality. At this point the team determined to work on coupling factor because strong coupling factor tends to lead to strong signal quality. The five factors (Factors, A, B, C, D, and E) highlighted in gray in Figure 13.2 are the factors the team determined to focus on.

13.5 Experimental Strategy

13.5.1 Response

Response for our experiment was a nominal-the-best type I. The item selected for the response was coupling factor. Coupling factor is a measure of the communication between the antenna in the vehicle and the transponder in the key.

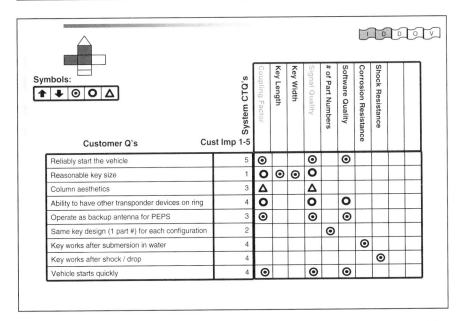

Figure 13.1   Customer Qs and system CTQs: The House of Quality Level 1

13.5.2 Noise Strategy

The noise strategy employed in this experiment was a simple one. This system is subjected to and influenced by its surrounding environment. The design of the surrounding trim can interfere with the signal and reduce the coupling factor.

We subjected the product to three levels of noise: no trim interference, mid level of trim interference, and a max level of trim interference that would subject the system to the most noise.

13.5.3 Control Factors

We determined to study one two level factor and four three level factors. Many orthogonal arrays could be chosen to conduct this experiment but the logical choice was the L18.

These control factors could easily be altered physically so a physical test was utilized.

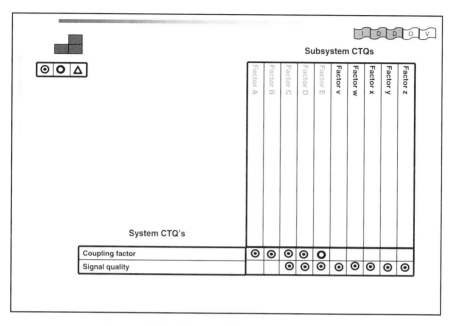

Figure 13.2 House of Quality: lower levels

13.5.4 Input Signal

Only one input signal is used by the vehicle so in our experiment only one input signal was utilized.

13.6 The System

Figure 13.3 illustrates the subsystem that is under study to optimize Factors A, B, C, D, and E to assure reliable communication between the key and the immobilizer module.

The coupling factor (Figure 13.4) between the vehicle mounted antenna and the transponder antenna depends upon the amount and direction of the magnetic field. The coupling factor is describing how many field lines of the vehicle mounted antenna are captured by the transponder antenna. In general, a high coupling factor is desired for reliable communication during authentication of the key.

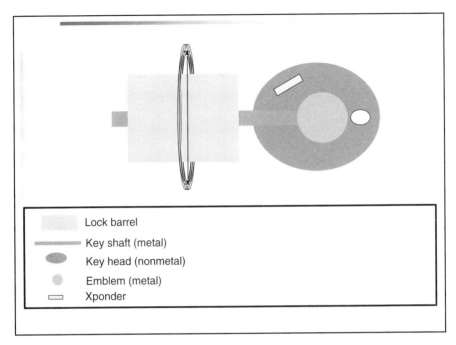

Figure 13.3 Diagram of subsystem

The coupling factor is a calculation based upon the voltage levels and inductance of the immobilizer antenna and transponder coil in the key (Figure 13.5). It is critical for voltage levels and inductance to be measured in the vehicle environment with all representative materials, especially metal.

13.7 The Experimental Results

The experiment was conducted and yielded the results as shown in Table 13.1.

Figure 13.6 shows the results of the response plots for the mean. These seem encouraging until compared to the response plots for the S/N Ratio (Figure 13.7). The numerous "V" shapes in these plots are somewhat alarming. They tend to imply interactions.

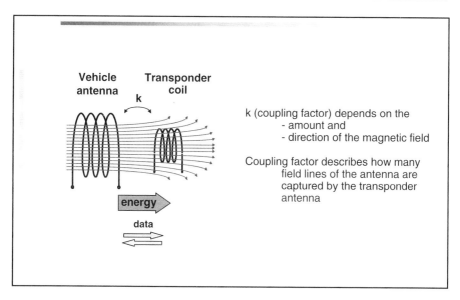

Figure 13.4 Coupling factor

The plot in Figure 13.8 shows the 18 actual test data points.

Using the information from the L18 orthogonal array, a full factorial was predicted and plotted as shown in Figure 13.9.

By restricting factor A to a value of -5, factor D to a value of 10 and factor B to a value of 47 we are left with the four values at the top of the chart shown in Figure 13.10. We then ran physical confirmation on all four of these configurations. All four systems confirmed within 10% with the actual values exceeding the predicted values.

13.8 Conclusions

This DFSS project resulted in predicting the values for the critical design factors to achieve an optimal transponder based immobilizer system design. This optimal design results in a 34% coupling factor improvement and 6.4 dB improvement in S/N ratio as compared to a current production design. Moreover, the parameters which determine high coupling factor are understood and we have gained the knowledge to enable first time quality in our engineering designs.

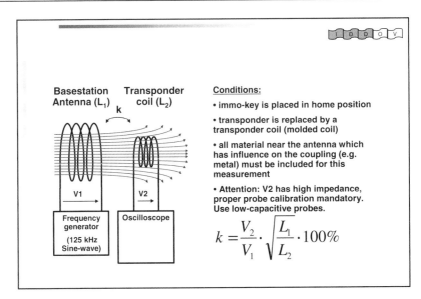

Figure 13.5   Measuring coupling factor

Table 13.1   Optimization details: L18 orthogonal array results

Run	Factor A	Factor B	Factor C	Factor D	Factor E	No Noise Factor I	Noise Factor 1	Noise Factor 2	σ	μ	S/N
1	-5	35	0	10	5	1.31253	1.30616	1.29675	0.01	1.31	44.32
2	-5	35	7.5	12	3.5	1.01376	1.00169	0.98322	0.02	1.00	36.25
3	-5	35	15	15	2	0.82857	0.85188	0.82766	0.01	0.84	35.69
4	-5	41	0	10	3.5	1.40523	1.30538	1.3484	0.05	1.35	28.63
5	-5	41	7.5	12	2	1.14747	1.15244	1.14863	0.00	1.15	52.91
6	-5	41	15	15	5	1.07813	1.07935	1.05702	0.01	1.07	38.63
7	-5	47	0	12	5	1.14971	1.15789	1.15764	0.00	1.16	47.90
8	-5	47	7.5	15	3.5	1.19877	1.19283	1.16416	0.02	1.19	36.13
9	-5	47	15	10	2	1.30758	1.3211	1.29616	0.01	1.31	40.41
10	-6.5	35	0	15	2	0.63595	0.63462	0.63174	0.00	0.63	49.40
11	-6.5	35	7.5	10	5	0.9195	0.91799	0.91824	0.00	0.92	61.11
12	-6.5	35	15	12	3.5	0.84225	0.84393	0.82755	0.01	0.84	39.37
13	-6.5	41	0	12	2	1.00826	0.9928	1.01279	0.01	1.00	39.63
14	-6.5	41	7.5	15	5	0.88834	0.89492	0.87091	0.01	0.88	37.06
15	-6.5	41	15	10	3.5	1.1336	1.13882	1.11154	0.01	1.13	37.83
16	-6.5	47	0	15	3.5	0.91679	0.90829	0.92807	0.01	0.92	39.32
17	-6.5	47	7.5	10	2	1.20464	1.20407	1.1923	0.01	1.20	44.73
18	-6.5	47	15	12	5	1.07811	1.07486	1.04615	0.02	1.07	35.65

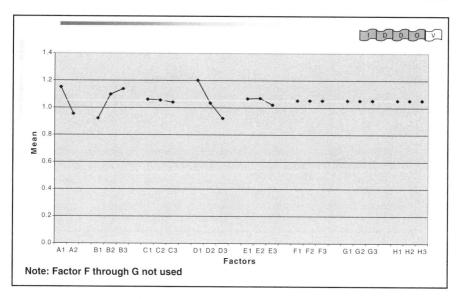

Figure 13.6 Optimization details: mean graphs

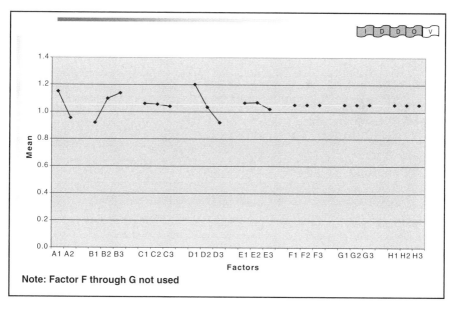

Figure 13.7 Optimization details: signal-to-noise graphs

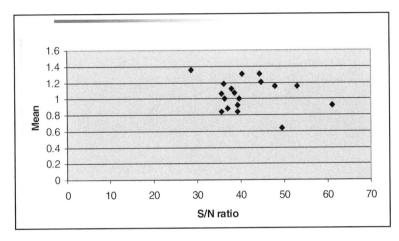

Figure 13.8   The original 18 experimental results

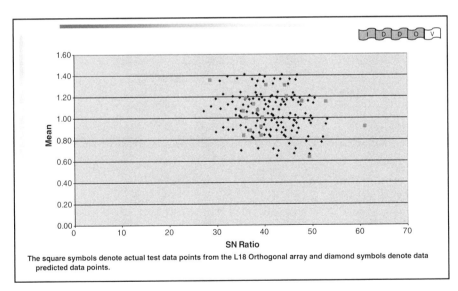

The square symbols denote actual test data points from the L18 Orthogonal array and diamond symbols denote data predicted data points.

Figure 13.9   Optimization details: based upon the L18 orthogonal array, a full factorial was generated and plotted

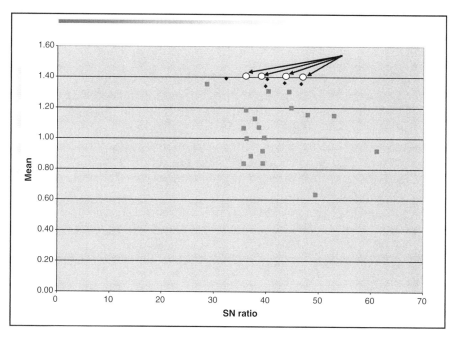

Figure 13.10 Optimization details: restricting factors A, B, and D

Another important learning from this project is that from time to time one may be confronted with systems filled with interactions. In these cases, one could reconsider the experimental set-up and seek an approach that eliminates or reduces the impact of these interactions. Another approach explored in this study was to utilize the information from the orthogonal array to predict the full factorial results and then attempt confirmation of the most interesting points identified in the analysis. This approach was successful in identifying combinations that otherwise were difficult to perceive.

13.8.1 Summary

Utilizing the Taguchi Robust Engineering (RE) approach as the foundation, we were able to develop an understanding of the critical parameters and their settings to improve coupling factor and reduce its variation in the face of noise.

13.8.2 Acknowledgments

We would like to thank the following individuals:

- David Trush and Natalie Wienckowski, DFSS team members from General Motors, for applying their specific expertise to this immobilizer project;
- Personnel from Philips Semiconductor Corporation for sharing their expertise in transponder based systems;
- Personnel from Robert Bosch Corporation for performing many test iterations and assuring accurate DFSS test results;
- Craig Jensen of ASI Consulting Group, LLC for his help and encouragement throughout this project;
- Matt Rudnick, of General Motors, for his assistance in data analysis.

This case study is contributed by William Biondo, Jill Griffin, of General Motors Corporation, USA.

Part Three
Subsystems Level Optimization by Suppliers

14

Magnetic Sensing System Optimization

ALPS Electric, Japan

14.1 Executive Summary

In recent years, the changing business environment has rapidly and significantly altered product development. The pace of change in the business environment was not imagined even a few years before. In traditional product development duplication of work and problems with material availability create waste and delays.

Our traditional product development process needed to be changed to compete in the new business environment. The way engineering work is conducted must mirror the changing business environment. For this reason our company adopted Quality Engineering as the method to achieve business success.

In order to create a new product development process, an example using Quality Engineering was created so that our engineers could easily understand the new approach. In the past, each subsystem was optimized one-by-one which was inefficient. In order to change the way of engineering work, it is necessary so generate a successful example, which our engineers can relate to and understand.

Robust Optimization: World's Best Practices for Developing Winning Vehicles,
First Edition. Subir Chowdhury and Shin Taguchi.
© 2016 Subir Chowdhury, Shin Taguchi, and ASI Consulting Group, LLC.
Published 2016 by John Wiley & Sons, Ltd.

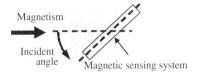

Figure 14.1   Principle of the magnetic sensing system

The "magnetic sensing system" was selected as a model example for demonstrating the new general purpose design techniques using Quality Engineering. For development of new magnetic sensing system, techniques and concepts suggested by quality engineering were applied completely.

Consequently, compared with the traditional development technique, a superior development process must be verified in terms of development cost and time to develop. As a result, the new engineering process based on Quality Engineering reduced development time and cost. Ultimately, we were able to change the way of engineering work aimed at establishment of a system design technique. The new techniques allow an entire system as a whole to be optimized rather than one subsystem at a time.

14.1.1   The Magnetic Sensing System

A magnetic sensing system is used to measure the angular displacement of an engine component in real time. The component's actual angular displacement is refered as the "incident angle" shown in Figure 14.1. A reference magnetic field corresponds to an incident angle of zero. When the component is rotated in relation the reference magnetic field, the sensor measures the changes in magnetic flux and computes the angular deflection. The magnetic sensing system is constructed with three subsystems as shown in Figure 14.2: a magnetic sensor, IC and software.

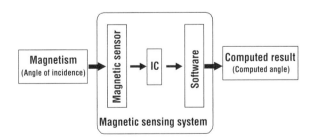

Figure 14.2   Construction of the magnetic sensing system

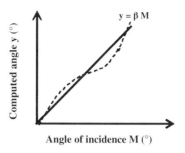

Figure 14.3   Ideal state

The basic function of this magnetic sensing system is specified as "Computing the actual angular deflection from the changes in the magnetic flux." The ideal state follows $y = \beta M$ a signal M is the incident angle or true angle and output y is computed angle or measured value, respectively (Figure 14.3).

However, the actual sensing system does not follow the ideal function. The actual system has variability and nonlinearity between the actual angle and the measured angle. Because sensitivity β is tunable by other factors, only the signal to noise ratio is applied in this optimization to an objective for optimization.

14.2   Improvement of Design Technique

14.2.1   Traditional Design Technique

In the traditional design technique, engineers who are assigned to respective subsystems specify subsystem parameter values according to the system specification. They will repeat simulation until specifications are met (Figure 14.4).

Because subsystems were developed separately, the performance specifications for those subsystems were set very conservatively. Many such specifications could not be easily achieved during the product development processes. These severely demanding specifications led to delays and higher costs. Even after releasing the subsystem designs, it was often difficult to manufacture the components, adding further delays and costs.

14.2.2   Design Technique by Quality Engineering

For the step-1 of 2-step optimization, it is important not to worry about meeting the target values of the system and subsystems, but to aim at maximizing the SN ratio.

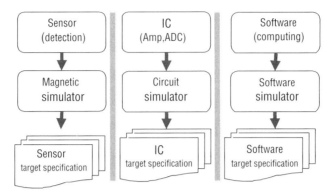

Figure 14.4 Traditional design technique

Using this system design flow, development target specifications of the respective subsystems are optimized in a state balanced for the entire system in consideration of relationship between the subsystems. Then, a modeling technique adopted in this time is explained. Modeling takes place in two great steps:

1. Relationship between input and output is determined using simulators.
2. A mathematical model expressing the relationship between input and output is prepared.

Preparation of this modeling formula is explained using the magnetic simulator as an example. First, basic features of input and output are determined from the magnetic simulator (Figure 14.5).

Mathematical formulas are formed using orthogonal polynomials [1] according to input and output features. Furthermore, a mathematical formula

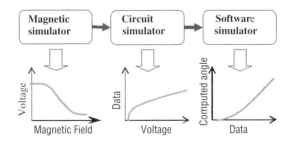

Figure 14.5 Technique on modeling

for magnetic field conversion is formed from a determinant in consideration of the rotation system. The final sensor model formulas are built up through combinations of several other formulas (Equations (14.1) and (14.2).

$$R = m + \beta_1(A - \overline{A})$$
$$+ \beta_2 \left[(A - \overline{A})^2 - \frac{k^2 - 1}{12} h \right]$$
$$+ \beta_3 \left[(A - \overline{A})^3 - \frac{3k^2 - 7}{20} h^2 \right] + \cdots \tag{14.1}$$

$$B = B_a \times \cos(\theta) \times \cos(\Gamma) \times \cos(\eta)$$
$$- \sin(\theta) \times \sin(\Gamma) + \cdots \cdots \cdots \cdots \tag{14.2}$$

In another circuit simulator and software simulator also, models are built up using a similar technique.

14.3 System Design Technique

14.3.1 Parameter Design Diagram

A parameter design diagram for the entire system is shown in Figure 14.6. Optimization was performed on the condition that the three subsystems were systematized to one, combining a sensor, IC and software. Candidates of control factors were reduced to 48 and assigned to L_{108}.

From the words of Dr. Taguchi, "The more complicated the control factors are, the better it is" and "Improvement can be performed according to the number of control factors," as many factors as possible were taken up and L_{108} was applied. Then, the biggest L_{108} in orthogonal tables of the mixed system familiar to us [2] was adopted.

Variability around the nominal value of 48 control factors were taken as noise factors. Variation of magnetic field strength was taken as a pure noise factor. Therefore, a total of noise 49 factors were assigned to an L108.

Variation of magnetic field strength was a noise factor that was uncontrollable by us. Therefore, a total of 49 noise factors were adopted in addition to the variability of 48 control factors.

For this system, it is clear that a consistent tendency of noise factors does not exist. Therefore, 49 noise factors were assigned to L_{108}.

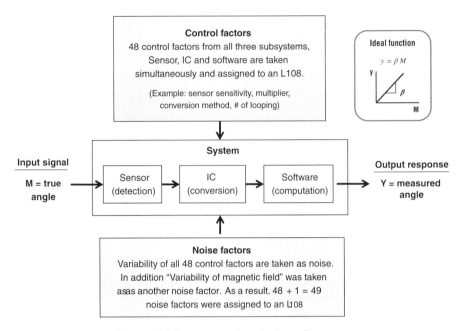

Figure 14.6 Parameter design diagram

14.3.2 Signal Factor, Control Factor, and Noise Factor

For the signal factors (Table 14.1), 360 levels were applied as input angles in the magnetic sensing system.

The control factors and the noise factors were set to ±5% and ±1% as shown respectively in Table 14.2 and Table 14.3. To reduce the effect of interaction between the control factors to the smallest possible, the level of the control factors was set as narrow as 5%. Computing instructions are shown in Table 14.4. An orthogonal table itself was constructed as linked to a simulation engine, and computing was performed automatically in only providing a level value according to an instruction of the direct product computing.

Table 14.1 Signal factor

M	M	M	M	⋯	M	M
Angle of incidence (°)	0	1	2	⋯	358	359

Table 14.2   Assignment of control factor (%)

Row	Control factor	Level 1	Level 2	Level 3
1	A: Sensor sensitivity 1	−5%	0%	+5%
2	B: Sensor sensitivity 2	−5%	0%	+5%
⋮	⋮	⋮	⋮	⋮
27	AA: IC, Voltage 1	−5%	0%	+5%
28	AB:IC Computing factor	−5%	0%	+5%
29	AC: IC Index	−5%	0%	+5%
⋮	⋮	⋮	⋮	⋮
48	AV: Software setting range	−5%	0%	+5%
49	e: Error row	−5%	0%	+5%

Table 14.3   Assignment of noise factor (%)

Column	Noise factor	Level 1	Level 2	Level 3
1	A: Error of sensitivity 1	−1%	0%	+1%
2	B: Error of sensitivity 2	−1%	0%	+1%
⋮	⋮	⋮	⋮	⋮
27	AA: Error of voltage 1	−1%	0%	+1%
28	AB: Error of computing factor	−1%	0%	+1%
29	AC: Error of index	−1%	0%	+1%
⋮	⋮	⋮	⋮	⋮
48	AV: Error of setting range	−1%	0%	+1%
49	AW: Strength of magnetic field	−1%	0%	+1%

Table 14.4   Robust of optimization layout $L_{108} \times L_{108}$

14.3.3 Implementation of Parameter Design

The initial stage of computing started immediately. A problem existed in which many incomputable rows were generated attributable to infeasible combination. The magnetic sensing system was entirely a newly developed article. For that reason, some of control factor combination of L108 went into infeasible range.

Many countermeasures were also studied and finally narrowed to one technique. The following technique was performed. The number of infeasible runs was counted and the average of SN ratios from feasible runs was computed for each factor level and graphed, as shown in Figure 14.7. From this result, the best level was identified as the level 2 of control factor for the next iteration. The computation became feasible for all 108 runs by the time of the 4th iteration.

By letting the 5th iteration as a new baseline sequential computing was further repeated [4] (Figure 14.8).

When it was judged that the SN ratio was not further improved after the 11th iteration, this condition was specified as an optimal condition.

14.3.4 Results of the Confirmation Experiment

The result of the 4th iteration where all 108 runs became feasible and the result of the 11th iteration as the optimum design are shown in Table 14.5. The db gain from the 4th to the 10th iteration is:

$$(-7.73) - (-46.48) = 38.75 \, (db)$$

Therefore, its great gain was confirmed.

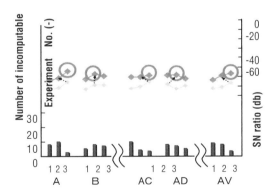

Figure 14.7 Effect of incomputable numbers and SN ratio

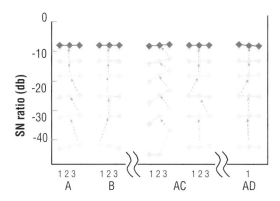

Figure 14.8 Response graph of SN ratio (sequential computing 5th iteration to 11th iteration)

Table 14.5 Results of the confirmation experiment (db)

	5th Iteration		11th & Last Iteration	
	Prediction	Confirmation	Prediction	Confirmation
Optimum	−30.18	−33.22	−6.05	−7.73
Baseline	−42.21	−46.48	−5.83	−6.52
Gain	12.03	13.26	0.98	1.21

The benefit is equivalent to reducing of the standard deviation to 1/86. Performance of the initial design and the optimum design are shown in Figure 14.9.

Results confirmed that reproducibility was sufficient. The amount of improvement through a series of computations was as high as about 39 decibels.

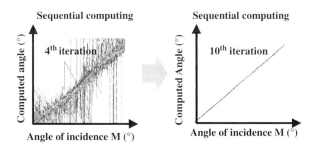

Figure 14.9 Comparison of confirmation experiment result

It is inferred that high reproducibility of benefits resulted from sequential iteration of computations. The great improvement was attributable to optimization, in which numerous control factors were applied.

Furthermore, when a prototype was produced at the optimal condition, as we anticipated, all requirements were met at once.

14.4 Effect by Shortening of Development Period

Improvements attributable to changing the traditional technique to Quality Engineering were estimated. In the case of the traditional development technique, the following problems existed. Numerous design changes resulted from poorly balanced specifications. For this problem, in the system design technique using Quality Engineering, the time to market was greatly shortened by having to develop a well balanced specification at the initial stage of development. The time to market is estimated as 1/3 or less compared with that of the traditional technique. It is estimated that reproduction of a prototype was greatly reduced; consequently, the cost for development became 1/5 or less.

Furthermore, because validity of a system specification can also be identified at the initial stage of development, the worst situation – that the development is cancelled at the last stage – can be avoided.

14.5 Conclusion

Through this series of activity, it can be said that "It was very satisfying to completely believe in the approach of Quality Engineering." It was felt that a stance at first to try an advised matter and then to discuss it was important especially in the case of the Quality Engineering.

Three individual effects are described.

For the first effect, it was clarified that the method in which optimization was performed on the entire local system from the beginning, had an advantage over other methods, in which a system was divided into subsystems and local optimization was performed, in quality and cost also. Secondly, it was clarified that drastic improvement can be achieved by assigning as many control factors as possible simultaneously to a large orthogonal array. Regarding db gain, it was confirmed that the amount of improvement became as high as about 39 decibels. Finally, as intended initially, we were able to verify this new form of engineering process. It can be said from the results of this optimization that three subsystems were optimized simultaneously. This was possible because of the use of powerful robust assessment by the outer array. It changed engineering work processes and an effective development technique was established. Development objectives were attained sufficiently.

This new way of engineering will be further expanded for in-house design and development.

14.5.1 Acknowledgments

Finally, we wish to express our gratitude to Mr. Hiroshi Yano and everybody of the NMS Society who guided us in advancing this study.

14.6 References

1. Taguchi, G. (1976) *Design of Experiments, Vol. 1 3rd edn.* Tokyo: Maruzen Co. Ltd.
2. Research Council for Simplification of Measuring Control in Metrology Administration Association: (1984) *Parameter Design in New-Product Development*, Japan Standards Association.
3. Yano, H. (1998) *The Guide to Quality Engineering Calculating Method.* Japan Standards Association.
4. Abe, M., Toyofuku, K. Fukunaga, S. and Watanabe, Y. (2005) Optimization of component characteristic for safety improvement against clash by simulation, *Journal of Quality Engineering*, **12**(4).

This case study is contributed by Yukimitsu Yamada, and Tomonari Ui of ALPS Electric, Japan.

15

Direct Injection Diesel Injector Optimization

Delphi Automotive Systems, Europe and Delphi Automotive Systems, USA

15.1 Executive Summary

Delphi Automotive Systems is entering the Direct Injection Diesel business, which requires a significant shift in technologies from the current diesel injection approach. The injector itself is the key to mastering this goal and its high sensitivity to sources of variation makes Robust Engineering a valuable approach to optimization.

A robust engineering dynamic experiment based on a numerical model of the injector allowed us to achieve the following:

- 4.46 dB (decibels) improvement in Signal to Noise ratio through parameter design. This represents a reduction of about 40% in the variability of injected quantity.
- An improvement of about 26% in the predicted manufacturing process end of line First Time Quality (% of good injectors at the end of the line).
- Generation of tolerance design based charts to support manufacturing tolerance decisions and further reduce cost.

Robust Optimization: World's Best Practices for Developing Winning Vehicles,
First Edition. Subir Chowdhury and Shin Taguchi.
© 2016 Subir Chowdhury, Shin Taguchi, and ASI Consulting Group, LLC.
Published 2016 by John Wiley & Sons, Ltd.

This Robust Engineering case study shows that *Cost* effective and *Informative* Robust Engineering projects can be conducted using good simulation models, with hardware being used to confirm results.

The case study also identified the need to extend the Tolerance Design method to dynamic response when a dynamic parameter design is used (not part of the current Taguchi Expert curriculum).

15.2 Introduction

15.2.1 Background

The Common Rail Direct Injection Diesel fuel system is an important technology for Delphi. For this reason, a product and process engineering team, dedicated to the design and implementation of the best possible common rail system with respect to both product and process, was set up at the European Technical Center. The Direct Injection Diesel common rail system is comprised of the following core components of the engine management system:

- injectors;
- high pressure pump;
- fuel rail and tubes;
- high pressure sensor;
- electronic control module;
- pressure control system.

The main challenge with diesel common rail systems as opposed to indirect diesel injection systems is the continuous high operating pressures. Current common rail systems are designed to operate with pressure levels of 1350 to 1600 bars.

Figure 15.1 shows a sample injector and its key variability sources. The injector is the most critical and the most complex element of the system. A Diesel Common Rail system will not be successful if the injectors are not "World Class" in term of quality and reliability.

15.2.2 Problem Statement

The injector complexity is due to very tight manufacturing tolerances and challenging customer requirements for injected quantity. These issues are confirmed by the problems reported by our competitors. Some of them

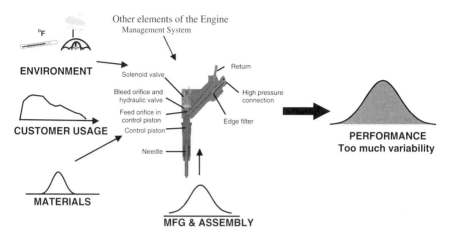

Figure 15.1 Injector and key variability sources

experienced very high scrap levels in production. An alternative approach is to build quality into the product at an early stage of design.

Injector operation is affected by the following sources of variability:

15.2.3 Objectives and Approach to Optimization

The optimization process followed the flow chart in Figure 15.2. A simulation model was developed, improved, and used to perform the orthogonal array experiments. Because of the very high confidence level in the simulation model (see Figure 15.3), we decided to use hardware only for the confirmation runs.

The main deliverables assigned to this optimization process were:

- Reduce the part-to-part and shot-to-shot variation in injected quantity.
  - Part-to-part variation is variation among several injectors.
  - Shot-to-shot variation is variation from one injection to the next within the same injector.
- Decrease sensitivity to manufacturing variation and be able to reduce cost by increasing component tolerances as appropriate.
- Provide graphical tools to increase understanding of downstream quality drivers in the design.

The labeling used in the experiment is shown in Figure 15.4.

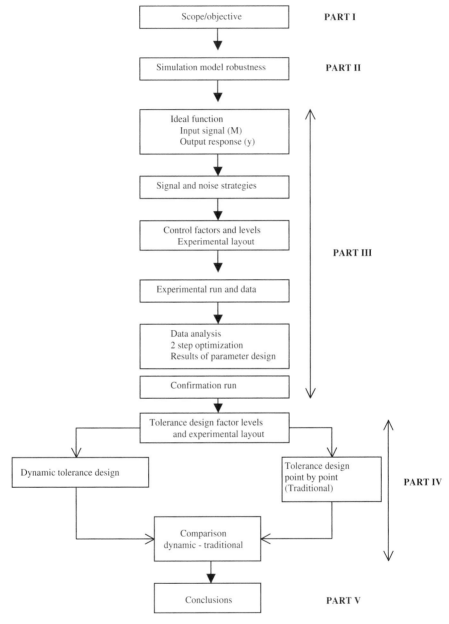

Figure 15.2 Optimization process flow chart

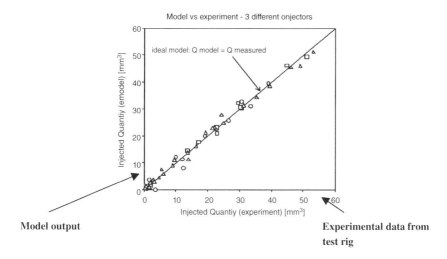

Model output

Experimental data from test rig

: reference - : control factor A,B,G,E, F changed - O: control factors A, B,G,E, I changed.

Figure 15.3 Comparison model data Vs experimental data for various rail pressure, pulse width and injector control factor setting

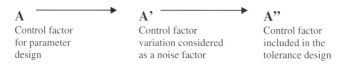

A
Control factor for parameter design

A'
Control factor variation considered as a noise factor

A"
Control factor included in the tolerance design

Figure 15.4 Labeling used in the experiment

15.3 Simulation Model Robustness

15.3.1 Background

A Direct Injection Diesel injector for a Common Rail system is a complex component, with high precision parts and very demanding specifications. To simulate such a component means representing the physical transient interactions of a coupled system including a magnetic actuator, fluid flows at very high pressure, possibly with cavitation phenomena, and moving mechanical parts. The typical time scale for operation is a few microseconds to a few milliseconds.

Because pressure wave propagation phenomena are very important in the accurate representation of injector operation, the simulation code cannot be

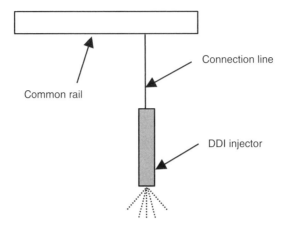

Figure 15.5 Modeled system

limited to the injector itself, but must also include the common rail and the connection pipe from the rail to the injector (Figure 15.5).

The Rail and the connection line are standard hydraulic elements, in which the model calculates wave propagation using classical methods for solving wave equations. The internal structure of the injector is illustrated on Figure 15.6.

We can distinguish two hydraulic circuits and one moving part (plus the mobile part of the control valve). The first hydraulic circuit feeds the control volume at the common rail high pressure through a calibrated orifice. The pressure P_C in the control volume is controlled by activating the electro-mechanical control valve and bleeding off a small amount of fluid. The duration of electrical activation, called pulse-width, is calculated by the Engine Control Module (ECM), depending upon driver demand, rail pressure and engine torque needs. The other hydraulic circuit feeds the nozzle volume at a pressure P_N that remains close to common rail pressure. Using internal sealing, the area stressed by the pressure is larger on the control volume side than on the nozzle side. Thus, as long as P_C and P_N remain equal, the needle is pushed against the nozzle seat and the injector is closed. To start an injection event, the control valve is activated, which lowers the pressure in the control chamber until the force balance on the needle changes sign. Then the needle moves up, and injection occurs. To close the injector, the control valve is closed and the pressure builds up again in the control chamber until the force balance on the needle changes sign again and the needle moves down and closes.

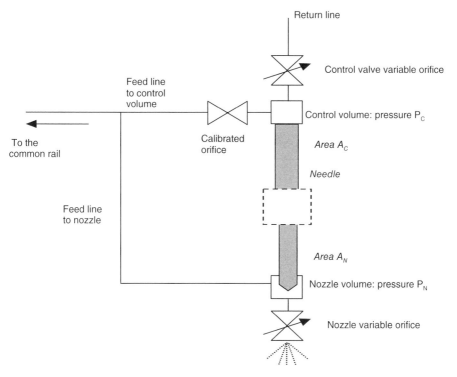

Figure 15.6 Schematic of the injector

The structure of the mathematical model of the injector can be deduced from its physical structure (Figure 15.7). Three submodels are coupled:

- electromechanical control valve;
- hydraulic circuits;
- mechanical moving parts.

Model inputs are:

- duration of electrical activation of electromagnetic actuator;
- rail pressure.

Model outputs include time variation of:

- magnetic force;
- fuel pressure and velocity in lines;

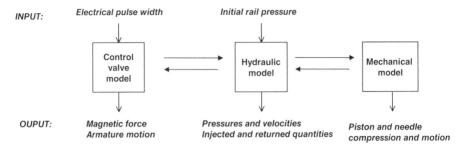

Figure 15.7 Structure of the mathematical model

- flows through orifices;
- displacement of moving parts;
- others . . .

The behavior of a Direct Injection Diesel injector can be characterized by a so-called "mapping," which gives the injected fuel quantity versus pulse-width and rail pressure (Figure 15.8).

Given the initial rail pressure and the pulse-width, the model is expected to accurately predict the quantity of fuel that should be injected.

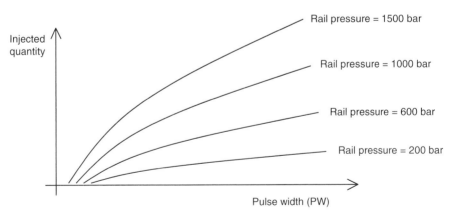

Figure 15.8 Schematic representation of injector mapping: Injected quantity *Vs* pulse-width for various rail pressure levels

15.3.2   Approach to Optimization

The difficulty in building and then optimizing such a model is that some features have a very strong effect on the results. Imprecision about their model representation can affect results in an unacceptable way. Experimental investigations have been used to thoroughly study the characteristics of internal flows in key parts like the control valve and the nozzle, for instance. From these experiments, empirical equations have been fitted and then implemented in the model. Finally, the transient nature of the flows is also a problem because most of the experiments are done statically.

In the end, after implementing the most realistic physical representation of variables, either based on experiments or on theoretical calculation, injector model optimization is a feedback loop between model results and experimentally observable injection parameters, like injected quantity, injection timing, control valve displacement. Engineering judgment is then necessary to assess the cause of any discrepancy and to improve or even change the physical representation of variables shown to be inadequate.

15.3.3   Results

Figure 15.3 shows the correlation between model and experiment for injected quantity, for rail pressures and pulse widths covering the full range of injector operation, for three different injectors with different settings of control factor values. The agreement between model and hardware is good over the full range of operation. Figure 15.3 *is the basis of the high confidence level we have in the model outputs.*

15.4   Parameter Design

15.4.1   Ideal Function

The Direct Injection Diesel injector has several performance characteristics:

- Recirculated quantity
- Total opening time
- Delay of main injection
- Injected quantity

In the traditional approach, these performance characteristics would require individual optimizations and trade-off based on experimental results.

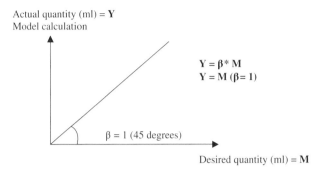

Figure 15.9   Ideal function

In such a case, the power of a carefully chosen ideal function is invaluable Figure 15.9. For an injection system, the customer requirement can be expressed as the quantity of fuel required for good combustion at a given operating point. By considering the customer need as a signal, we can compute through the simulation model, under noise conditions, what the actual injector will deliver. This last value is the y-axis of the ideal function. A perfect injector will have a slope of one, with the actual quantity equal to the desired value.

A design that exhibits the above ideal function is likely to be acceptable for any of the performance characteristics listed above.

15.4.2 Signal and Noise Strategies

15.4.2.1 Signal Levels

Five levels of desired injected quantity were selected based on customer specifications:
2 ml, 5 ml, 15 ml, 25 ml and 40 ml.

15.4.2.2 Noise Strategy

Table 15.1 shows the impact of the noise factors on the injected quantity. A (+) indicates that an increase in the noise factor will increase the injected quantity (direct relationship). A (−) indicates that a decrease in the noise factor will increase the injected quantity (inverse relationship).

The compounded noise level 1 (N1) groups all the noise factor levels that have the effect of reducing the injected quantity. The compounded noise level 2 (N2) groups the noise factor levels that have the effect of increasing the injected quantity.

Table 15.1 List of noise factors

Noise factor	Level 1	Level 2	Relationship with injected quantity	Compounded noise level 1 (N1)	Compounded noise level 2 (N2)	Rationale
B'	−10	+ 10	+	Level 1	Level 2	Mfg
C'	−5	+ 5	+	Level 1	Level 2	Age
D'	−2	+ 2	+	Level 1	Level 2	Mfg, Age
E'	−3	+ 3	-	Level 2	Level 1	Mfg, Age
F'	−3	+ 3	-	Level 2	Level 1	Age
G'	−3	+ 3	+	Level 1	Level 2	Mfg, Age
H'	−4	+ 4	-	Level 2	Level 1	Age
I'	−30	+ 30	+	Level 1	Level 1	Mfg
J'	−3	+ 3	+	Level 1	Level 2	Age
L	0.6	1	+	Level 1	Level 2	Mfg, Age
M	−10	+ 10	+	Level 1	Level 2	System
N	0	0,03	+	Level 1	Level 2	Mfg
O	2	5	+	Level 1	Level 2	Mfg

Mfg = Manufacturing variation; Age = Aging; System = Influence of other components.

B', C' . . . J' are noise factors obtained by considering variation (from manufacturing or other sources) of the corresponding control factors. L, M, N and O are noise factors not derived from control factors. We expect these noises to be present during product usage.

15.4.3 Control Factors and Levels

Control factors and levels are listed in Table 15.2.

An $L_{27(3^{13})}$ orthogonal array was used to perform the experiment. Two columns were unused. Figure 15.10 shows the parameter design.

15.4.4 Experimental Layout

Table 15.3 shows the experimental layout.

15.4.5 Data Analysis and Two-Step Optimization

The data was analyzed using the dynamic signal to noise ratio with the below formula:

$$\eta = 10 \log \frac{\frac{1}{r}(S_\beta - V_e)}{V_e}$$

Table 15.2 Control factors list and levels

Factor	Level 1	Level 2	Level 3
A	−16.6 %	**X**	+ 16.6%
B	**−11.1%**	X	+ 11.1%
C	−16.6 %	X	+16.6%
D	**−11.1%**	X	+ 11.1%
E	−28.6%	X	**+ 28.6%**
F	−18.5%	**X**	+ 18.5%
G	−18.2%	X	**+ 22.7%**
H	+ 33.3%	**X**	+ 66.6%
I	−11.1%	X	**+ 11.1%**
J	−8.62%	X	**+ 6.9%**
K	−2.3%	X	**+ 2.3%**

X is a reference value; current design levels are in bold type.

S_β = Sum of squares of distance between zero and the least square best fit line (Forced through zero) for each data point.

Ve = Mean square(Variance); r = Sum of squares of the signals.

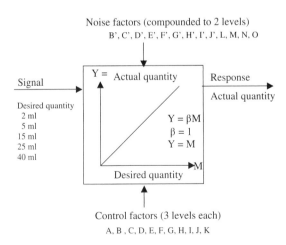

Figure 15.10 Parameter design diagram

Table 15.3 Experimental layout

Control factor array														Signal	2 ml		5 ml		15 ml		25 ml		40 ml		S/N	β
#	A	B	C	D	E	F	G	H	I	J	K	12	13	Noises	N1	N2	N1	N2	N1	N2	N1	N2	N1	N2		
1	1	1	1	1	1	1	1	1	1	1	1	1	1													
2	1	1	1	1	2	2	2	2	2	2	2	2	2													
3	1	1	1	1	3	3	3	3	3	3	3	3	3													
25	3	3	2	1	1	3	2	3	2	1	2	1	3													
26	3	3	2	1	2	1	3	1	3	2	3	2	1													
27	3	3	2	1	3	2	1	2	1	3	1	3	2													

DATA

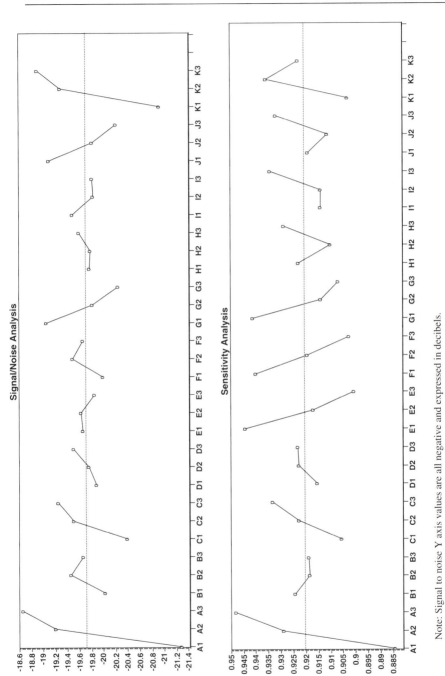

Note: Signal to noise Y axis values are all negative and expressed in decibels.

Figure 15.11   Signal to noise ratio and sensitivity plots

Table 15.4 Two-step optimization

Factors	A	B	C	D	E	F	G	H	I	J	K
Initial design	A2	B1	C1	D1	E3	F2	G3	H2	I3	J3	K3
S/N maximization	A3	B2	C3	D3	E?	F2	G1	H3	I1	J1	K3
Adjustment for beta					E1						
Optimized design	A3	B2	C3	D3	E1	F2	G1	H3	I1	J1	K3

1st step

2nd step

Design change required   Design change required

Table 15.5 Prediction

	Signal/noise (dB)	β (Slope)
Initial design	−20.7118	0.8941
Optimized design	−15.0439	1.0062
Gain	+ 5.66	0.1121

Figure 15.11 shows S/N ratio and sensitivity plots. Table 15.4 shows two-step optimization and Table 15.5 shows the prediction.

15.4.6 Confirmation

Table 15.6 shows the model confirmation.

The model confirms the expected improvement with a slight difference. Neither the optimum nor the current design combinations were part of the control factor orthogonal array. The initial hardware testing on the optimized configuration shows promising results. The hardware confirmation will be completed before end of 1999. Figure 15.12 shows graphical results initial design compared to optimized design for parameter design.

Table 15.6 Model confirmation

	Predicted		Model confirmation		Hardware confirmation	
	S/N (dB)	Beta	S/N (dB)	Beta	S/N	Beta
Initial design	−20.7118	0.8941	−20.2644	0.9150	Ongoing	Ongoing
Optimum design	−15.0439	1.0062	−15.8048	0.9752	Ongoing	Ongoing
Gain	5.66	0.1121	4.46	0.06		

15.4.7 Discussions on Parameter Design Results

15.4.7.1 Technical

- B and C are control valve parameters. The design change on these parameters suggested by parameter design is likely to improve injected quantity, part to part, and shot to shot variation. The simulation model confirms these improvements. Hardware confirmation is ongoing.
- A, D, E, G, H, I, J are hydraulic parameters. Implementing the changes suggested by parameter design will decrease our sensitivity to most of the noise factors and to pressure fluctuations in the system.

15.4.7.2 Economical

An improvement in signal to noise ratio can be directly translated into a reduction in variability.

$$Variability\ improvement = (1/2)(\hat{Gain}/6) * initial\ variability$$

Gain = signal to noise ratio gain in decibels\hbox{\curr;} in our case, the model confirmed gain is 4.46 dB (see Table 15.6). We are making an almost 40% reduction in variability from the initial design.

The slope of the ideal function is distributed normally [2] (Table 15.7). The tolerance on the slope can be obtained from the tolerances at each signal point of injection (Figure 15.13).

From the tolerances on β, and assuming a normal distribution, we can draw a curve to relate end of line scrap level or First Time Quality (FTQ) to the deviation from the target β. This is not meant to reject the loss function approach. (We use the loss function in tolerance design, Section 15.5.). Rather it is to quantify the reduction in part rejection (and therefore cost avoidance) at the end of the assembly line if we implement the recommendations from

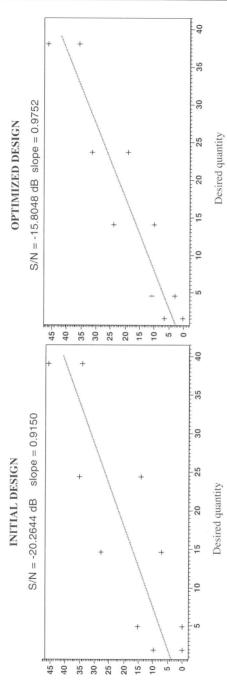

Figure 15.12 Graphical results initial design compared to optimized design for parameter design

Table 15.7 Specifications on slope

Desired quantity	Minimum acceptable quantity (at 3 sigma)	Target	Maximum acceptable quantity (at 3 sigma)
2	1	2	3
25	21.5	25	28.5
40	36.5	40	43.5
β (slope)	0.945 (at 3 sigma)	1	1.054 (at 3 sigma)

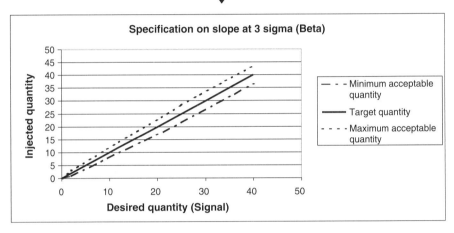

Figure 15.13 Slope tolerance: graphical view

parameter design. The end of assembly line First Time Quality (FTQ) is calculated by the following formula:

$$FTQ = (Number\ of\ injectors\ within\ tolerances)/(Total\ number\ of\ injectors\ produced)$$

Figure 15.14 identifies FTQ.

If β is the observed slope, $\mu = 1$ is the mean, z is the number of standard deviations away from the mean, $\phi\ (z)$ is the standard cumulative normal distribution function,

$$\phi(z) = [1 \div (2\pi)^2] * \int_{-\infty}^{z} e^{[-(x^2)\div 2]}dx$$

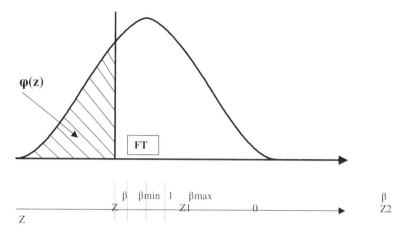

Figure 15.14 FTQ representation

with

$z = (x - \mu)/\sigma$
σ is the observed standard deviation on β
σ_i is the standard deviation on β before parameter design
σ_{opt} is the standard deviation after parameter design
$\beta_{min} = 0{,}945$
$\beta_{max} = 1{,}054$
$Z2 = (\beta_{max} - 1)/\sigma$ and $Z1 = (\beta_{min} - 1)/\sigma$

$$FTQ = \phi(Z2) - \phi(Z1) = [1 - \phi(Z1)] - \phi(Z1) = 1 - 2\phi(Z1) = 1 - 2\phi[(\beta_{max} - 1)/\sigma]$$

So FTQ is a function of the observed standard deviation of β.

If the standard deviation of β from the initial design is σ_i, considering the 40% reduction in variability from parameter design, the standard deviation of β from the optimized design after parameter design is

$$\sigma_{opt} = \sqrt{0.6 * \sigma_i^2} = 0.77 * \sigma_i$$

The predicted end of line fraction of good parts is obtained through the curve in Figure 15.15. We will simulate the improvement on FTQ with two different values of σ_i on β. It is hard to know the current value of σ_i, but we know that the optimized standard deviation is $(0.77 * \sigma_i)$.

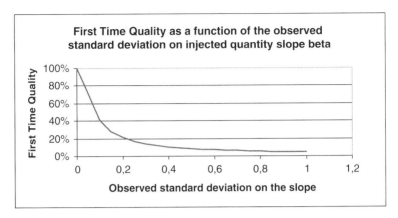

Figure 15.15   End of line FTQ as a function of the observed standard deviation on β

Table 15.8   End of line FTQ as a function of the standard deviation on slope

σi	σopt	FTQi	FTQopt	ΔFTQ %	% gain (on FTQ %) initial to optimized
0.10	0.076	41%	52%	11%	26%
0.05	0.038	72%	84%	12%	16%

Possible FTQ improvement in % with a standard deviation on β from the initial design equal to σ_i. (Table 15.8).

15.5 Tolerance Design

Tolerance design is traditionally conducted with nondynamic responses. With a dynamic robust experiment, this is not optimal and can result in endless trade off discussions. The ranked sensitivity to tolerances might be different from one signal point to another. A way to solve this is the use of "dynamic tolerance design." In this robust design experiment, tolerance design was performed both on a signal point basis and a dynamic basis. In dynamic tolerance design, optimization is based on β instead of the injected quantity. The tolerance on β is obtained by considering the tolerances at each signal point and drawing a best-fit line. This ensures a more comprehensive tolerance design (Figure 15.16).

To illustrate, we will perform tolerance design with the two approaches and compare them.

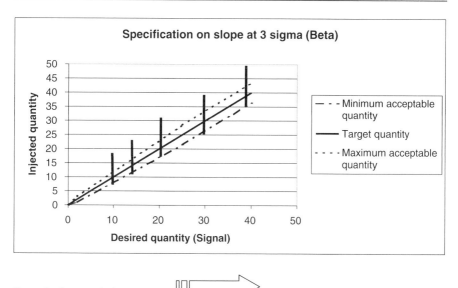

Several tolerance designs ⟹ One dynamic tolerance design

Figure 15.16   Tolerance design: signal point Vs dynamic

15.5.1   Signal Point by Signal Point Tolerance Design

15.5.1.1   Factors and Experimental Layout

We are using a 3-level tolerance design (Table 15.9). We have 13 parameters in the tolerance design study. Therefore, we will once again use an **L** 27.

X is the optimized level of each design parameter; σ is the current tolerance of each design parameter (Table 15.10).

The signal point based tolerance design is performed using the data in the corresponding signal point column.

15.5.1.2   Analysis of Variance (ANOVA) for Each Injection Point

Table 15.11 shows ANOVA signal point 1 and Figure 15.17 shows percent contribution of design factors to variability per signal point.

15.5.1.3   Loss Function

For the signal point tolerance design, the loss function is nominal-the-best on injected quantity. Table 15.12 shows percent contribution of design factors to variability. This table was performed for each signal point. Figure 15.18 illustrates the Taguchi loss function.

Table 15.9 Factors and levels for tolerance design

(Nominal Value is the Optimum Level from Parameter Design)

Tolerance Factors	Sigma (%)	Level-1	Level-2	Level-3
A"	3.810	Nominal$-\sqrt{3/2}\times\sigma$	Nominal	Nominal$+\sqrt{3/2}\times\sigma$
B"	3.704	:	:	:
C"	2.381	:	:	:
D"	1.333	:	:	:
E"	0.010	:	:	:
F"	0.004	:	:	:
G"	0.002	:	:	:
H"	2.667	:	:	:
I"	6.250	:	:	:
J"	0.002	:	:	:
K"	0.114	:	:	:
L	0.022	:	:	:
M	2.632	:	:	:

Table 15.13 gives a summary of current total loss for some signal points. Figure 15.19 is intended to support manufacturing tolerance decision-making.

15.5.2 Dynamic Tolerance Design

The control factor settings are the same as in the signal point tolerance design. In this case, the response becomes the β from each combination. We use a

Table 15.10 Tolerance design experimental layout

	Tolerance Factors Assigned to L27													M=Signal, y=Response				
No	A"	B"	C"	D"	E"	F"	G"	H"	I"	J"	K"	L	M	M1 2ml	M2 5ml	M3 15ml	M4 25ml	M5 40ml
1	1	1	1	1	1	1	1	1	1	1	1	1	1	y1	y2	y3	y4	y5
2	2	2	2	2	2	2	2	2	2	2	2	2	2	y6	y7	y8	y9	y10
3	3	3	3	3	3	3	3	3	3	3	3	3	3	y11	y12	y13	y14	y15
:	:	:	:	:	:	:	:	:	:	:	:	:	:	:	:	:	:	:
:	:	:	:	:	:	:	:	:	:	:	:	:	:	:	:	:	:	:
:	:	:	:	:	:	:	:	:	:	:	:	:	:	:	:	:	:	:
:	:	:	:	:	:	:	:	:	:	:	:	:	:	:	:	:	:	:
25	3	3	2	1	1	3	2	3	2	1	2	1	3	y121	y122	y123	y124	y125
26	3	3	2	1	2	1	3	1	3	2	3	2	1	y126	y127	y128	y129	y130
27	3	3	2	1	3	2	1	2	1	3	1	3	2	y131	y132	y133	y134	y135

Table 15.11 ANOVA signal point 1

Source	DF	S	V	F	S'	ρ (%)
A″	2	2.6413	1.3206	15.9851	2.4760	12.63%
B″	2	4.3410	2.1705	26.2721	4.1758	21.30%
C″	2	0.0672	0.0336			
D″	2	2.2450	1.1225	13.5872	2.0798	10.61%
E″	2	0.3953	0.1977			
F″	2	0.9158	0.4579	5.5426	0.7506	3.83%
G″	2	1.0480	0.5240	6.3427	0.8828	4.50%
H″	2	0.0332	0.0166			
I″	2	0.6072	0.3036	3.6750	0.4420	2.25%
J″	2	0.5165	0.2582	3.1257	0.3512	1.79%
K″	2	2.0346	1.0173	12.3134	1.8693	9.53%
L	2	4.2696	2.1348	25.8403	4.1044	20.93%
M	2	0.4940	0.2470	2.9899	0.3288	1.68%
e1						
(e)	6	0.4957	0.0826		2.1480	10.95%
Total	26	19.6087	0.7542			100.00%

Injection point 1: 2 ml desired (the ANOVA was performed at each signal point).

nominal-the-best approach with the target $\beta = 1$ (ideal function). Table 15.14 shows the experimental layout for dynamic tolerance design.

15.5.2.1 Dynamic Analysis of Variance

β replaces the injected quantity. The target for β is 1 in this case. The boundaries for injector First Time Quality correspond here to the upper

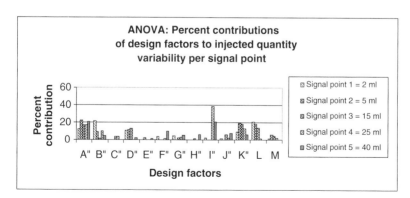

Figure 15.17 Percent contribution of design factors to variability per signal point

Table 15.12 Percent contribution of design factors to variability, signal point 1: 2 ml desired

Factor	ρ (%)	Ltc	Current σpc	Lcp
A″	12.63	6.9235 C	3.81%	0.8744381 C
B″	21.30	6.9235 C	3.70%	1.474706 C
C″	0	6.9235 C	2.38%	0
D″	10.61	6.9235 C	1.33%	0.7345834 C
E″	0	6.9235 C	0.01%	0
F″	3.83	6.9235 C	0.00%	0.2651701 C
G″	4.5	6.9235 C	0.00%	0.3115575 C
H″	0	6.9235 C	2.67%	0
I″	2.25	6.9235 C	6.25%	0.1557788 C
J″	1.79	6.9235 C	0.00%	0.1239307 C
K″	9.53	6.9235 C	0.11%	0.6598096 C
L	20.93	6.9235 C	0.02%	1.449089 C
M	1.68	6.9235 C	2.63%	0.1163148 C

ρ % is the percentage contribution of the design parameter to the injected quantity variability;
Ltc is the total current loss in $ in the Taguchi loss function;
σpc is the design parameter current standard deviation or its current tolerance level;
σt is the current total output standard deviation, considering variation of all design parameters;
σp is a variable representing the standard deviation of a design parameter;
Lcp is the current fraction of the total loss caused by the corresponding parameter; the higher the
% contribution (%r), the higher the Lcp for a given design parameter;
FTQ is First Time Quality (end of assembly line);
Lp is the loss caused by a design parameter, given any value of the parameter standard deviation
σp (Lp is a function of σp);
K = proportional constant of the loss function;
$K = Ao/\Delta o^2 = 9.18$ C;
Ao = base average injector cost in $ = $C;
Δo = specification on injected quantity = 0,33 ml at 1σ;
VT = current total variance from ANOVA = 0.7542;
Ltc = $K * VT = $ 6.9235*C.

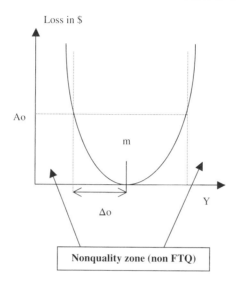

Figure 15.18   Taguchi loss function

and lower lines we can draw on each signal point based on customer specifications (see Figure 15.13). Table 15.15 gives the data for dynamic tolerance design ANOVA.

15.5.2.2 Dynamic Loss Function

Comparison Dynamic Tolerance Design – Signal Point Tolerance Design

The total loss with dynamic tolerance design is almost half of the total loss when tolerance design is conducted signal point by signal point. We do not have a precise explanation for this finding. The percent contributions of the design parameters at each signal point (see Figure 15.17) are difficult to use for decision making, due to the complexity of trade-off. We found it easier to

Table 15.13   Summary of current total loss for some signal points

Signal point	Current total loss
2 ml	$ 6.9235 C
25 ml	$ 0.5541 C
40 ml	$ 0.5541 C

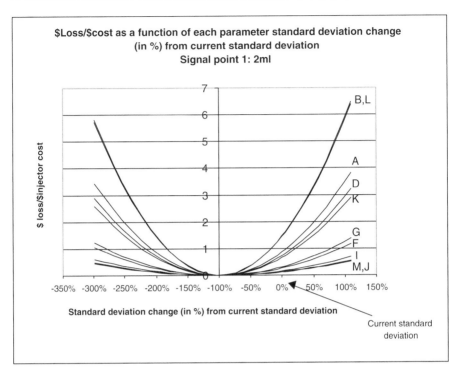

Figure 15.19 Signal point 1:2 ml; chart to support manufacturing tolerance decision-making

Table 15.14 Experimental layout for dynamic tolerance design

#	A"	B"	C"	D"	E"	F"	G"	H"	I"	J"	K"	L	M	
	Tolerance factors													
1	1	1	1	1	1	1	1	1	1	1	1	1	1	$\beta 1$
2	1	1	1	1	2	2	2	2	2	2	2	2	2	$\beta 2$
3	1	1	1	1	3	3	3	3	3	3	3	3	3	$\beta 3$
25	3	3	2	1	1	3	2	3	2	1	2	1	3	$\beta 25$
26	3	3	2	1	2	1	3	1	3	2	3	2	1	$\beta 26$
27	3	3	2	1	3	2	1	2	1	3	1	3	2	$\beta 27$

Table 15.15 Dynamic tolerance design ANOVA

Source	DF	S	V	F	S'	ρ (%)
A″	2	0.0117	0.0059	35.4725	0.0114	35.50%
B″	2	0.0010	0.0005	2.9348	0.0006	1.99%
C″	2	0.0066	0.0033	20.1535	0.0063	19.73%
D″	2	0.0001	0.0000			
E″	2	0.0006	0.0003			
F″	2	0.0002	0.0001			
G″	2	0.0000	0.0000			
H″	2	0.0005	0.0003			
I″	2	0.0029	0.0015	8.9340	0.0026	8.17%
J″	2	0.0038	0.0019	11.5096	0.0035	10.82%
K″	2	0.0005	0.0003			
L	2	0.0004	0.0002			
M	2	0.0037	0.0018	11.0911	0.0033	10.39%
e1						
e2						
(e)	14	0.0023	0.0002		0.0043	13.39%
Total	26	0.0320	0.0012			100.00%

consider the "global percent contributions" obtained from dynamic tolerance design (see Figure 15.20).

15.6 Conclusions

15.6.1 Project Related

a. A gain of 4.46 dB in signal to noise ratio is realized in the modeled performance of the Direct Injection Diesel injector. This gain represents a reduction of about 40% in injected quantity variability. We anticipate that the variability reduction will translate to a 16–26% increase of the fraction of parts inside injected quantity tolerances at the end of the manufacturing line.

b. Several charts were developed to support manufacturing process tolerance decision making. As information on tolerance upgrade or degrade cost is made available, the charts will be used to further reduce cost by considering larger tolerances where appropriate.

Table 15.16 Total current loss distribution among parameters

Factor	ρ (%)	Ltc	Lcp
A″	35.5	3 C	1.065 C
B″	1.99	3 C	0.0597 C
C″	19.73	3 C	0.5919 C
D″	0	3 C	0
E″	0	3 C	0
F″	0	3 C	0
G″	0	3 C	0
H″	0	3 C	0
I″	8.17	3 C	0.2451C
J″	10.82	3 C	0.3246 C
K″	0	3 C	0
L	0	3 C	0
M	10.39	3 C	0.3117 C

Dynamic characteristic: slope β.
Same as signal point by signal point tolerance design. Response is β in this case.

ρ is the percent contribution of the design parameter on the β variability;
K = poportional constant of the loss function;
$K = Ao/\Delta o^2 = 2500 * C$;
Ao = average injector cost in \$ = \$ C;
Δo = specification on β = 0.02 at 1σ;
VT = current total variance from ANOVA = 0.0012;
Ltc = **K** ∗ **VT** = \$ 3∗C.

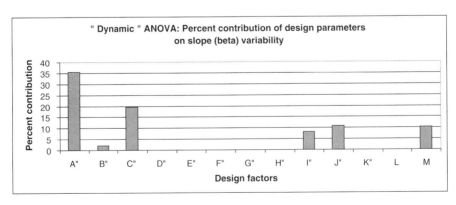

Figure 15.20 Percent contribution of design factors on the slope (dynamic tolerance design)

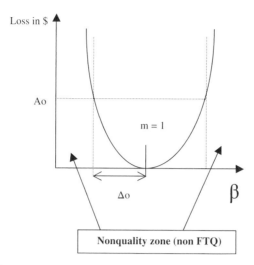

Figure 15.21   Taguchi loss function for dynamic tolerance design

15.6.2 Recommendations for Taguchi Methods

a. We recommend the use of dynamic tolerance design for dynamic responses. Optimization at individual signal points may not always lead to a clear and consistent design recommendation. This case study proposes a solution for a nominal-the-best situation for tolerance design. A standardized approach needs to be developed to cover dynamic tolerance design.

b. Hardware cost is one of the roadblocks to Taguchi Robust design implementation. The development of good simulation models will help overcome this roadblock while increasing the amount of good information one can extract from the projects. With a good model, we can conduct parameter design and tolerance design with a great number of control factors. We are also able to simulate control factor manufacturing variation as noise factors. The ability to consider manufacturing variation is an important advantage compared to full hardware based robust engineering. Hardware confirmation should be conducted to ascertain the results and should not lead to a surprise if a good robustness level is achieved up-front on the simulation model.

15.6.3 Acknowledgments

We wish to thank Dr Jean Botti, Dr Martin Knopf, Aparicio Gomez, Mike Seino, Carilee Cole, Giulio Ricci from Delphi Automotive Systems; and Shin Taguchi, Alan Wu from the ASI Consulting Group, LLC, USA for their technical support.

15.7 Reference and Further Reading

1. Wu, Y. and Wu, A. (1998) *Taguchi Methods For Robust Design*, New York: ASME Press.
2. Neter, J. (1983) *Applied Linear Regression Models*. Homewood, IL: Richard D. Irwin, Inc.

This case study is contributed by Desire Djomani and Pierre Barthelet of Delphi Automotive Systems, Europe and Michael Holbrook of Delphi Automotive Systems, USA.

16

General Purpose Actuator Robust Assessment and Benchmark Study

Robert Bosch, LLC, USA

16.1 Executive Summary

Intake manifolds incorporate electronic actuators to operate air control valves for tuning and emission purposes. This benchmark study determines the robustness and performance levels of eight competitive actuators using the Taguchi optimization methods.

Relevant performance characteristics of the actuators are current draw, torque output, angle of rotation, and response time. The ideal function includes these parameters with an input signal of electrical power (voltage x current) and an output response of mechanical power (torque x angle/time). Two test temperatures and four different voltages create noise and signal conditions.

Robust Optimization: World's Best Practices for Developing Winning Vehicles,
First Edition. Subir Chowdhury and Shin Taguchi.
© 2016 Subir Chowdhury, Shin Taguchi, and ASI Consulting Group, LLC.
Published 2016 by John Wiley & Sons, Ltd.

The benchmark testing necessitated a practical, low-cost test setup capable of mounting the various actuators, providing a consistent torque load, and measuring the parameters. This testing used an existing torque versus angle measurement tool with minor modifications. A set of clock springs produced the torque loads needed. These springs, recovered from consumer grade tape measures, were inexpensive. They provide nearly constant torque over 120 degrees of actuation and can be wound/unwound to produce the necessary torque. The torque levels were established with zero, one, two, or three springs in parallel.

The robust assessment approach to benchmarking performance generated results that separated the eight different designs by measuring the robustness of the base functional characteristics that are important to the customer.

16.2 Introduction

As the demand for engine efficiency increases, automakers turn to devices, such as air control valves in the intake manifold and intake runners, to reduce emissions. Electronic solenoids and actuators are replacing the traditional vacuum solenoids for powering and controlling these air valves. A performance benchmark study is performed as part of an investigation into making such an electronic actuator more competitive.

Traditional benchmark approaches measure individual performance criteria under standard test conditions. This practice does not provide a realistic assessment of the product's performance under operating conditions because it ignores robustness. Additionally, it may require several tests to identify all criteria. This study simplifies the benchmark process by examining the performance of the system's core "ideal" functions using a single test procedure.

The robustness and performance levels of eight competitive actuators are tested using equipment that identifies power draw and mechanical output. These two functions incorporate the important customer measures: operating voltage, current draw, actuation time, torque capabilities, and actuation angle. To keep the test procedure simple and low-cost, a series of consumer tape measure springs provide the mechanical torque resistance for the actuators to rotate against.

16.3 Objectives

A performance benchmark study of electronic actuators is necessary as part of a product's competitive evaluation. The test method used does not follow the traditional Bosch approach, which would determine several specific and individual functions or symptomatic failures. Instead, the Taguchi Robust Engineering Method, developed by Dr. Genichi Taguchi, is used to assess actuator performance robustness against usage noise conditions. Separation of the actuator designs is achieved by examining the performance of the system's core "ideal" functions in the presence of noise conditions. The measure used is the signal-to-noise ratio.

The nature of this investigation requires a simple benchmarking test that reveals the most about the performance of the actuators' functions important to the customer. The Taguchi Method can compare the benchmark actuators with the use of a single, properly designed test that measures the ideal function against usages or noise conditions. The benchmarking method therefore follows the first six steps of Dr. Taguchi's eight-step optimization process to identify the robustness of the actuators.

16.3.1 Robust Assessment Measurement Method

16.3.1.1 Test Equipment

Benchmark testing requires the capability to determine the input and output energy of the ideal function. The available small-motor dynamometers are only effective for output shafts that spin quickly with multiple rotations. The cogging effect of these dynamometers produce irregular torque loads for the benchmark actuators, which turn less than full rotations and therefore were not useful in this investigation. Using an existing test setup, available in the timeframe of this investigation was preferred, and a search for an adaptable test bench returned a throttle body spring test fixture. This adaptation was consistent with the spirit of Dr. Taguchi's theme of fast and simple testing to show robustness.

An existing spring torque test bench measures torque and angle in the range of the benchmark actuators. It consists of a fixture and a data acquisition unit that measures the torque difference between two shafts. One end of the shaft connects to the measured device and the other end to a torque load. Modifications include a fixture plate to mount the actuators, shaft couplers to adapt each actuator's output shaft, a fixture to produce torque loads, and changes to the data acquisition software to record current and actuation time measurements. Figure 16.1 shows the test setup including the original spring test unit in the center.

The mounting plate is built with positioned holes to fasten all of the test actuators at their intended mounting location. The legs are adjustable in height to align the output shaft with the test fixture. Shaft adapters, built with

Figure 16.1   Benchmarking test setup

an SLS rapid prototype process, connect each actuator's output shaft to the round test fixture shaft. The adapters are built from SLS material because the process is fast, with an available in-house prototype lab, and the material is durable enough for the testing extent of this investigation.

The fixture to the right of the data acquisition unit in Figure 16.1 produces the torque load. The test plan requires that the torque load against which the actuator rotates is constant through the output shaft's travel, and that the torque load is adjustable to four different intensities. Considered for this task were: hydraulic rotary dampers, magnetic dampers, weights on pulleys, and tension springs on pulleys. Rotary dampers, particularly adjustable magnetic dampers, can produce constant torque loads but are costly. Furthermore, they were not available during the timeframe of this investigation. Weights are not effective because the high inertial forces from an actuator's fast response produce nonconstant torque loads. Tension springs are also not effective because their force increases with extension.

Constant force "clock" springs are used in the benchmark test fixture because their torque loads remain fairly constant with limited rotational displacement. Up to three springs are combined in parallel to produce the necessary torque test ranges. Figure 16.2 shows the torque load fixture with two springs. The springs were taken from consumer grade retracting tape measures purchased at a local hardware store. They are inexpensive and can rotate about 45 turns. The fixture clamps the clock spring housings, preventing them from rotating, and a shaft attached to the measurement unit holds

Figure 16.2   Torque load fixture

the center of the springs. The shaft has a slot that clamps the spring and winds it when turned.

Figure 16.3 shows the relationship between the number of rotations that the springs are wound and the correlating torque load. The four configurations

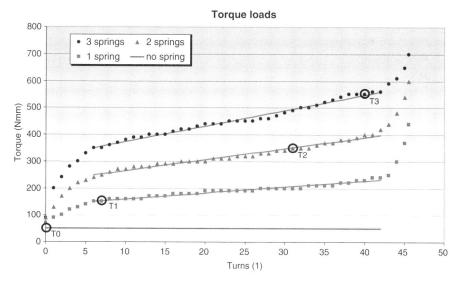

Figure 16.3   Tape measure spring windings to torque relationship

are: no spring, 1 spring, 2 springs, and 3 springs. These measurements were made with a handheld torque watch at the test stand shaft where the actuator output shaft attaches, to determine the number of windings necessary to achieve varied load conditions for benchmark testing. The torque watch was rotated against the load so that frictional forces in the test stand are included in the torque load measurements.

Four different torque load strengths are used during benchmark testing to identify the performance range of the actuators. The first, T0, is without any springs, and generates a torque load of about 0.05 Nm from the friction in the test system. The T1 condition is with one spring wound seven times, generating a torque load of about 0.15 Nm. The T2 condition with two springs wound 31 times generates a 0.30 Nm torque load. The T3 condition with 3 springs wound 40 times generates a 0.55 Nm torque load.

16.3.1.2 Data Acquisition

The data acquisition software for the spring torque tester was modified to record two additional channels for current and time, in addition to the original torque versus angle measurements. The channels measure voltage differences; therefore to measure current draw, an ammeter shunt is wired in series with the actuator ground wires. The voltage drop across the shunt is proportional to the current flowing through it. The test stand uses a 10 A/5 mV shunt, which means that a 10 amps draw creates a 5 mV voltage drop. An amplifier multiplies the voltage signal one hundred times so that the measurement channel can accurately detect the small voltage differences. The software divides the amplified voltage measurement by 0.5 to record the actuator current draw.

The data acquisition system records a measurement every 0.08789 degrees of rotation but does not record a time measurement. The computer's internal clock is not precise enough to record fast measurements and retrofitting the measurement equipment is costly. As an alternative solution, an external generator is used that transmits a voltage signal for the data acquisition software to interpret. The signal generator produces the burst function shown in Figure 16.4. With a 20 V amplitude and 500 mHz frequency, the voltage signal increases 10 V per second. The data acquisition software measures the voltage value and multiplies it by 0.1 to record the time stamp of each individual data point. Actuation time is calculated from the time stamps and corresponding angle measurements. The generated signal may not be precise where the function drops off, and therefore time stamps are measured during the function's increasing slope. When actuators take longer than

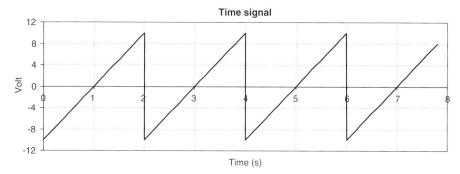

Figure 16.4 Generated voltage signal for actuation time measurement

2 seconds to actuate, a signal frequency of 250 mHz and a multiplication factor of 0.2 are used, as it stretches the increasing slope over 4 seconds.

Voltage to the actuator is supplied with a regulated power supply. It provides the controlled signal input to the actuator system and is used directly as part of the input energy. The four voltage signal inputs used for testing are near the operating range of the actuator. The highest signal input, labeled V4, is regulated at a 15 V supply. The V3 signal input is regulated at 12 V and the V2 signal input is regulated at 10 V. The signal input labeled Vmin is the minimum voltage that permits full actuation of the output shaft and is therefore unique to each actuator.

The measurements are recorded during the time that an actuator moves from the default position to the actuated position. No measurements are recorded during an actuator's return motion. The four test runs Vmin to V4 take less than one minute at any torque load. Changing the torque loads can take several minutes when it is necessary to add and wind a spring.

16.3.1.3 Data Analysis Strategy

The data acquisition software creates a data file for each test run that includes the time stamp, torque, and current draw as a function of angle. An Excel workbook calculates the input signal and output response of the actuator system. It also calculates actuation angle, actuation time, maximum rotational velocity, startup current, average torque load, and generates an angle versus time graph, current versus time graph, and torque load versus angle graph.

16.4 Robust Assessment

This project followed the robust engineering process to assess the performance of the benchmark actuators without optimizing the designs. The assessment provides key market understanding and function performance goals.

16.4.1 Scope and P-Diagram

The scope of the ideal function is the conversion of electrical power into the system to mechanical power out of the system. The system analyzed during benchmarking is the manifold air valve actuator as pictured in Figure 16.5. A test procedure for a complete actuator most accurately simulates customer applications and does not alter interactions between subsystems. The P-diagram in Figure 16.6 summarizes the parameters that define the robust assessment of the manifold air valve actuator. The diagram shows the system's input signal, different benchmark actuators, noise factors that cause variability, failure modes, and system output response.

16.4.2 Ideal Function

The ideal function of the actuator system is to convert electrical energy from the vehicle's battery into mechanical energy to position the air control valve. Figure 16.7 represents the ideal function in terms of power conversion. The electrical power is the signal (M) supplied to the actuator. The mechanical power is the response (y), since actuation time and torque are dependent on the electrical power input.

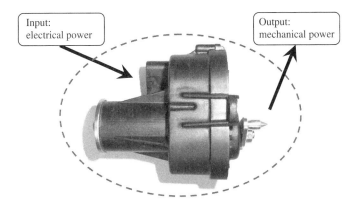

Figure 16.5 Scope

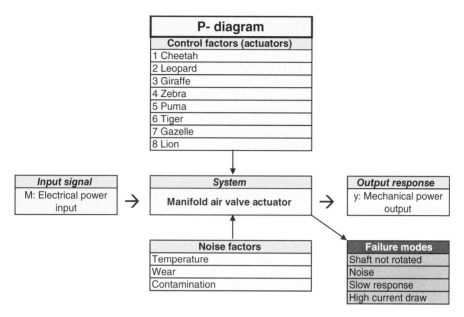

Figure 16.6 P-diagram

Beta (β), the slope of the ideal function, represents the efficiency of tested actuators. A steeper slope, and therefore a higher beta value, indicates that a system has been tuned to actuate against stronger torques in less time per given electrical power input.

The input signal is calculated from the controlled voltage, the current draw measurements, and the time measurements. Electrical power is voltage multiplied by current, where current is the electrical charge carried through a system in a specific timeframe. Current draw varies with time during actuation; therefore the power input calculation requires the integration shown in Equation (16.1). The integral of this equation, represented by the

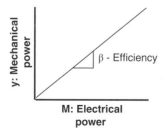

Figure 16.7 Ideal function

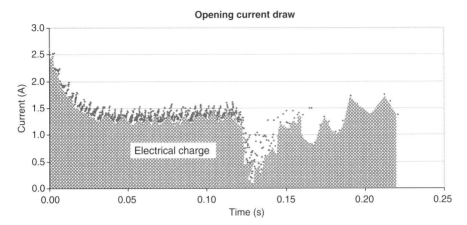

Figure 16.8 Representation of electrical charge calculation

shaded area in Figure 16.8, calculates the electrical charge through the system from the measured current draw and time data. Electrical charge multiplied by the controlled voltage is the electrical energy, which, divided by actuation time, is the input power.

$$M = P_{in}[W] = \frac{V \cdot \int_{0}^{t_{max}} i \cdot dt}{t} \tag{16.1}$$

The output response is calculated from the torque load, actuation angle, and actuation time measurements. Mechanical power for a rotating system is torque, multiplied by the rotation angle, divided by time. The clock spring configuration determines the approximate torque load on the system, but the actual load on the output shaft can vary due to friction within the clock springs. Therefore, the torque load is measured during actuation travel to determine the real power output. Since the load varies, the power calculation requires the integration shown in Equation (16.2). The integral of this equation, represented by the shaded area in Figure 16.9, calculates the work of the system from the measured torque and actuation angle data. The work divided by the actuation time is the mechanical power output.

$$y = P_{out}[W] = \frac{\int_{0}^{\theta_{max}} \tau \cdot d\theta}{t} \tag{16.2}$$

This study uses the Taguchi comparison metric signal-to-noise ratio to determine the robustness of the actuators. First, the focus is on measuring

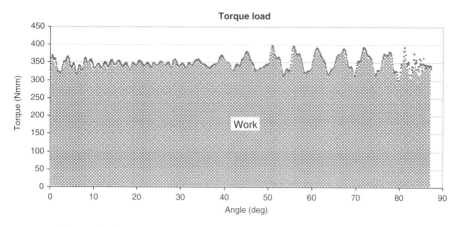

Figure 16.9 Representation of mechanical work calculation

robustness using signal-to-noise (S/N) and then on measuring performance efficiency using beta (β). The S/N of each actuator is calculated using the "Dynamic Most Common Case" formulation as outlined below in Equation (16.3). Variance is estimated by Equation (16.4), using Equations (16.5), (16.6) and (16.7) where n is the number of tests, M is the input signal data, and y is the output response data. The beta of the ideal function is equivalent to the efficiency of the actuator system, which is calculated using Equation (16.8).

$$S/N = 10 \log \left(\left(\frac{1}{r} \right) \left(\frac{(S_\beta - V_e)}{V_e} \right) \right) \tag{16.3}$$

$$V_e = \frac{S_T - S_\beta}{n - 1} \tag{16.4}$$

$$S_T = \sum_{i=1}^{n} y_i^2 \tag{16.5}$$

$$S_\beta = \frac{\left(\sum_{i=1}^{n} M_i y_i \right)^2}{r} \tag{16.6}$$

$$r = \sum_{i=1}^{n} M_i^2 \tag{16.7}$$

$$\beta = \sum_{i=1}^{n} \frac{M_i y_i}{r} \tag{16.8}$$

16.4.3 Signal and Noise Strategy

The electrical power in and the mechanical power out of the actuator are functions of voltage, current, torque, actuation angle, and actuation time as Equations (16.9) and (16.10) show.

$$P_{in} = V \cdot i \qquad\qquad (16.9)$$

$$P_{out} = \frac{\tau \cdot \theta}{t} \qquad\qquad (16.10)$$

The input signal for Equation (16.9) is determined by controlling the voltage and measuring current draw and actuation time. The actuators are tested at four voltages labeled Vmin to V4 to determine the slope of the ideal function. The current flow to the actuator is not controlled during testing because the actuator determines current draw in customer applications.

Torque output, actuation angle, and actuation time determine the output response for Equation (16.10). An actuator's maximum torque output can only be measured against a fixed load cell. This limits the measurements to a single output shaft position, identifying stall torque instead of actuation perform-ance indicative of real customer usage. This study measures performance of the entire actuation range by applying a constant torque load against which the output shaft rotates. Four different strength torque loads, labeled T0 to T3 are used during testing to generate a range of measurements that identify actuator performance capabilities.

Taguchi describes noise factors as "uncontrollable variables that affect the system's function response." Noises that vary the performance of a manifold air valve actuator include temperature, extended wear, misalignment, and contamination. These factors exist in all applications, and an ideal actuator produces the same output function under all conditions. A robust design is one that best approaches this ideal state. The benchmarking study is limited to temperature as a noise condition to induce variability in the results. This differentiates the actuators and makes it possible to determine which is most robust.

The first series of tests labeled N1 are conducted with the actuators at room temperature. Afterwards, the second series of tests labeled N2 are conducted with the actuators at an elevated temperature. For the N2 noise condition, the benchmark actuators are placed in an industrial oven at the elevated temper-ature to allow the temperature to soak into the part.

Table 16.1 shows the 32 configuration outer array. There are four torque load intensities from T0 to T3. For each torque load, an actuator is tested with

Table 16.1   Outer array

	N1 Temperature				N2 Temperature			
	T0	**T1**	**T2**	**T3**	**T0**	**T1**	**T2**	**T3**
V4	T0 V4	T1 V4	T2 V4	T3 V4	T0 V4	T1 V4	T2 V4	T3 V4
V3	T0 V3	T1 V3	T2 V3	T3 V3	T0 V3	T1 V3	T2 V3	T3 V3
V2	T0 V2	T1 V2	T2 V2	T3 V2	T0 V2	T1 V2	T2 V2	T3 V2
Vmin	T0 Vmin	T1 Vmin	T2 Vmin	T3 Vmin	T0 Vmin	T1 Vmin	T2 Vmin	T3 Vmin

four voltages from Vmin to V4. The table also shows that tests are conducted at room temperature and repeated at the elevated temperature to induce noise. This is known as a predominant noise strategy.

16.4.4   Control Factors

This benchmarking study tests and compares eight different actuators listed under control factors in the P-Diagram (Figure 16.6).

16.4.5   Raw Data

The signal (M) and response (y) values (calculated from the raw data using Equations (16.1) and (16.2) described in Section 16.4.2) are tabulated in Table 16.2. The results for each actuator are used to calculate the Dynamic signal-to-noise ratio (S/N).

16.4.6   Data Analysis

Figures 16.10 and 16.11 show the performance benchmarking results with the mechanical output response (y) as a function of the electrical signal input (M). The lines through the data points in the figures represent the zero point proportional betas, which are equivalent to each system's efficiency. A steeper slope, and therefore a larger beta value, indicates a more efficient system.

The proximity of the data points to the beta line indicates the robustness of the system. Closer points correspond to less variance in the design, which indicates that the system will perform more consistently. The signal-to-noise ratio captures the variability of the system's performance proportional to the output.

Table 16.3 summarizes the robust assessment results. It includes each actuator's beta value and signal-to-noise ratio, for which a higher positive value indicates less variability in the system. The market shows a wide range of robustness with a 42 dB spread. The "Lion" actuator is the most robust,

Table 16.2 Performance test input signal and output response

		N1 Temperature								N2 Temperature							
		T0		T1		T2		T3		T0		T1		T2		T3	
		M	y	M	y	M	y	M	y	M	y	M	y	M	y	M	y
Cheetah	V4	10.66	1.21	14.09	2.41	19.99	4.62	27.71	5.99	12.54	1.12	15.42	2.37	21.78	4.03	29.69	4.61
	V3	8.44	0.96	10.95	1.96	16.49	3.47	21.77	3.80	9.53	0.88	11.40	1.80	17.11	2.93	22.95	2.21
	V2	6.96	0.80	9.09	1.58	13.35	2.59	17.78	2.22	7.45	0.69	9.60	1.38	13.88	1.91	22.70	0.00
	Vmin	4.64	0.45	5.91	0.83	8.66	0.77	15.72	0.72	4.73	0.36	6.07	0.61	10.51	0.21	20.79	0.90
Leopard	V4	16.17	0.63	19.54	1.62	30.69	1.88	37.53	2.11	17.20	0.69	21.92	1.84	34.04	1.40	37.54	2.79
	V3	12.26	0.48	15.66	1.29	22.91	1.57	28.85	1.30	13.40	0.50	16.80	1.21	25.74	1.03	30.00	0.00
	V2	10.54	0.35	13.02	0.94	18.19	1.31	25.20	0.00	10.93	0.35	13.83	0.75	21.50	0.00	21.50	0.00
	Vmin	6.78	0.14	8.84	0.24	17.34	0.69	26.23	1.01	7.13	0.10	11.31	0.32	22.41	0.72	35.60	1.65
Giraffe	V4	51.08	0.85	55.35	0.00	55.35	0.00	55.35	0.00	42.41	0.48	43.95	0.00	43.95	0.00	43.95	0.00
	V3	33.37	0.40	35.76	0.00	35.76	0.00	35.76	0.00	27.20	0.00	28.08	0.00	28.08	0.00	28.08	0.00
	V2	23.88	0.02	25.10	0.00	25.10	0.00	25.10	0.00	19.50	0.00	19.50	0.00	19.50	0.00	19.50	0.00
	Vmin	61.22	0.82	97.20	0.00	97.20	0.00	97.20	0.00	51.73	0.40	76.40	0.00	76.40	0.00	76.40	0.00
Zebra	V4	55.77	0.82	60.07	2.40	63.74	1.84	72.75	0.00	47.90	0.82	50.77	1.88	55.40	0.54	59.25	0.00
	V3	33.64	0.77	36.05	1.49	42.97	0.88	46.80	0.00	30.66	0.72	33.28	1.18	33.98	0.04	38.16	0.00
	V2	23.14	0.54	25.98	1.07	31.99	0.02	32.60	0.00	21.45	0.45	23.41	0.60	26.30	0.00	26.30	0.00
	Vmin	19.08	0.46	24.48	1.01	109.15	3.60	124.14	0.10	21.07	0.46	27.77	0.91	95.63	2.36	101.20	0.00

Puma	V4	7.74	0.80	11.06	1.77	17.43	3.04	24.70	3.68	6.96	0.82	10.10	1.81	17.20	2.84	26.40	2.80
	V3	4.87	0.63	7.86	1.36	13.23	2.16	19.59	2.17	4.80	0.61	7.18	1.36	10.82	1.72	14.59	1.85
	V2	3.35	0.51	6.15	1.06	10.96	1.50	15.91	1.18	3.22	0.50	5.80	1.06	8.45	1.19	17.40	0.00
	Vmin	2.22	0.34	4.97	0.87	11.02	1.55	20.96	2.45	2.84	0.45	5.41	1.03	12.96	1.91	23.52	2.52
Tiger	V4	9.35	0.63	14.38	1.37	21.09	1.80	30.67	1.59	10.62	0.65	13.82	1.39	21.58	1.48	28.09	0.89
	V3	6.29	0.53	10.72	0.95	17.22	1.16	22.78	1.03	7.16	0.51	10.26	0.85	16.27	0.87	22.33	0.16
	V2	6.57	0.26	8.60	0.67	12.51	0.80	17.96	0.23	4.57	0.36	7.33	0.64	12.39	0.55	18.80	0.00
	Vmin	3.28	0.20	5.27	0.40	9.59	0.37	25.95	1.07	2.98	0.20	5.07	0.35	9.73	0.14	24.81	0.63
Gazelle	V4	11.48	1.28	18.26	2.72	58.80	0.00	58.80	0.00	13.28	1.13	16.07	2.10	36.15	0.00	36.15	0.00
	V3	7.65	0.89	9.30	1.50	31.32	0.00	31.32	0.00	8.07	0.79	11.15	1.16	25.92	0.00	25.92	0.00
	V2	5.66	0.69	9.51	1.24	17.50	0.00	17.50	0.00	5.54	0.62	8.81	0.93	18.10	0.00	18.10	0.00
	Vmin	3.35	0.49	5.17	0.94	16.83	1.04	106.80	0.00	3.72	0.36	5.59	0.64	77.80	0.00	77.80	0.00
Lion	V4	7.50	0.33	9.79	0.92	13.61	1.77	17.77	2.14	8.08	0.34	9.31	0.95	12.52	1.66	15.45	1.95
	V3	5.46	0.24	6.88	0.70	10.15	1.36	13.38	1.54	5.25	0.26	7.04	0.70	9.22	1.21	15.33	1.33
	V2	3.60	0.21	5.02	0.57	7.74	1.06	11.58	1.08	3.42	0.21	4.83	0.57	6.97	0.93	12.35	0.93
	Vmin	2.48	0.15	3.54	0.39	5.80	0.64	10.21	0.50	2.15	0.15	3.48	0.37	6.17	0.57	12.41	1.20

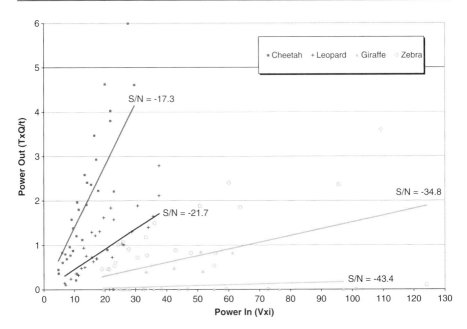

Figure 16.10   Robust Assessment (actuators 1–4)

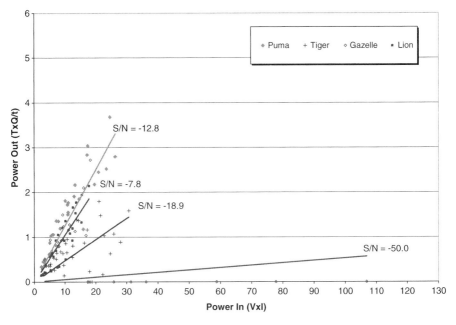

Figure 16.11   Robust Assessment (actuators 5–8)

Table 16.3 Dynamic signal-to-noise results

Assessment summary			
Unit	β	S/N	Rank
Cheetah	0.139	−17.3	3
Leopard	0.045	−21.7	5
Giraffe	0.002	−43.4	7
Zebra	0.015	−34.8	6
Puma	0.125	−12.8	2
Tiger	0.047	−18.9	4
Gazelle	0.005	−50.0	8
Lion	0.104	−7.8	1

with a 5.0 dB gain over the second-ranked "Puma" actuator and a 42.2 dB gain over the eighth-ranked "Gazelle" actuator. The "Lion" actuator is 16.8% less efficient than the "Puma" actuator, but with the 5 dB gain, equivalent to a variability reduction of about 40%. This shows that the "Lion" actuator is a better design for applications that require torque capabilities in the tested range from 0.05 nm to 0.55 Nm. It performs more consistently in the temperature noise condition while being adequately efficient.

The "Giraffe," "Zebra," and "Gazelle" actuators have the lowest S/N and beta values, partly because they failed to rotate at the higher torque loads. The "Giraffe" actuator can not rotate against the T1, T2, and T3 torque loads at 0.15 Nm, 0.35 Nm, and 0.55 Nm respectively. The "Zebra" and "Gazelle" actuators can not actuate against the T2 and T3 torque loads. These actuators could potentially perform well in applications where the torque requirement is below 0.35 Nm.

The intended torque capabilities of the actuators are unknown; therefore the robust assessment analysis was repeated without including the higher torque load data, to assess robustness at lower torques. Table 16.4 summarizes the robust assessment results using only data from the lower T0 and T1 torque load tests.

S/N is higher for all models in the lower torque assessment. The "Lion" actuator, with the least variability for all tested torque loads, was comparatively less robust at lower torque applications, ranking fourth. Notable is the improvement of the "Gazelle" actuator from eighth rank to third, indicating that it may not be intended for higher torque loads, but that it is a robust design in applications in which the operating torque is not higher than

Table 16.4 Low torque dynamic signal-to-noise results

Assessment summary			
Unit	β	S/N	Rank
Cheetah	0.136	−7.7	5
Leopard	0.058	−15.0	6
Giraffe	0.004	−38.2	8
Zebra	0.029	−22.3	7
Puma	0.159	−1.0	1
Tiger	0.082	−4.2	2
Gazelle	0.124	−5.8	3
Lion	0.081	−6.4	4

0.15 Nm. The "Tiger" actuator is also more robust in the lower torque range compared to the entire torque range, ranking second instead of fourth.

The "Puma" actuator is the best overall actuator for multiple applications. It is most efficient in the lower torque applications and second most efficient in all tested torque applications. More importantly, it has the least variability at the lower torque applications and second least variability for all tested torque applications.

16.5 Conclusion

The success of this project is measured by the objectives outlined in Section 16.3. The robust assessment approach to benchmarking performance generated results that separated the eight different actuator designs by measuring the robustness of the base functional characteristics that are important to the customer. The method distinguished the designs on a more fundamental level than traditional benchmarking practices which focus on specific characteristics. The benchmarking investigation benefited from the robust assessment approach, because in addition to determining performance behavior, the method indicated how consistently the products perform under customer usage.

The temperature noise strategy effectively produced variation in the output response to determine the most robust design. Although there is a large spread in the variability of the designs, the results are not an all-encompassing representation of customer usage, since the investigation was limited to only one noise condition. Future assessments would benefit from the introduction of additional noise factors such as wear from extended operation.

The test method identified relevant performances using a single test setup and procedure. The measurement results demonstrate that inexpensive consumer grade tape measure springs are an effective and practical way to apply a constant torque load over a limited rotation range. They can be wound/unwound and combined in parallel to produce the necessary torque. As opposed to friction devices, the torque load from springs exists even before the output shaft begins to rotate. The data used for the robust assessment was also useful for understanding other characteristics of the benchmark actuators. By analyzing the curves of the output shaft motion and current draw, for example, it was possible to gather information about each actuator's operation and control strategy.

Performing the benchmark investigation by analyzing the ideal function of the actuator resulted in a more valid comparison true to customer usage. The assessment showed that is possible to compare actuator performance using a simple test. This will impact future test practices as it proved the value and cost benefit of focusing on the measures related to the systems "core" ideal function.

16.5.1 Acknowledgments

I would like to acknowledge and extend my sincere gratitude to the following persons for their valuable time and assistance:

- Craig Jensen from ASI Consulting Group LLC, USA for sharing his wisdom, expertise, and enthusiasm;
- John Casari for mentoring me in the Robust Engineering process and assisting me with this chapter;
- Paul Rossi for providing me with the opportunity to develop this project;
- Blake Jarvis for his assistance with this project;
- Toan Tran for his technical support on the data acquisition software.

16.6 Further Reading

1. ASI Consulting Group LLC, *Design for Six Sigma (DFSS) Bosch Project Leader Training Week B*, Revision 2.4, Copyright 2006.
2. Chowdhury, Subir (2002). *Design For Six Sigma*. Chicago: Dearborn Trade Publishing.

This case study is contributed by Michael Neumeyer of Robert Bosch LLC, USA.

17

Optimization of a Discrete Floating MOS Gate Driver

Delphi-Delco Electronic Systems, USA

17.1 Executive Summary

A Floating MOSFET Gate Drive (or "Boot-Strap" circuit) is a commonly used circuit in automotive electronics design. Taguchi Methods were used to optimize the performance of a discrete circuit that could replace an existing IC (Integrated Circuit) solution with significant cost reduction. An initial concept was developed but this exhibited severe variation over the intended operating range, and as such was not yet capable of the intended application. The control factors chosen were based on important component values and parameters, but also on several different subcircuit topologies.

The application of Taguchi Methods was successful in reducing variability over the operating range, improving absolute performance, and also indicating important component and subcircuit interactions that in some cases were counter-intuitive. The ability to evaluate several circuit topologies and identify ultimate circuit tendencies greatly reduced design time over previous methods and resulted in a drastically improved design.

Robust Optimization: World's Best Practices for Developing Winning Vehicles,
First Edition. Subir Chowdhury and Shin Taguchi.
© 2016 Subir Chowdhury, Shin Taguchi, and ASI Consulting Group, LLC.
Published 2016 by John Wiley & Sons, Ltd.

17.2 Background

One of the major functions of an ECM (Engine Control Module) is the control of several types of solenoid such as fuel injectors. A simple solenoid consists of several turns (or loops) of wire around a bar of ferrous metal. By passing a current through the wire a magnetic field is induced which will force the bar to move in the direction of the rotating magnetic field. Reversing the direction of current flow can reverse the direction of movement. Typically the bar will be spring-loaded and controlling the average current through the solenoid in a single direction changes the relative position.

The magnetic properties of a solenoid mean that the current does not flow instantaneously like in a resistor, but instead rises and falls at a rate determined by the opposing magnetic fields. The average current is controlled by the PWM (Pulse Width Modulation) of a switch, commonly a Power MOSFET. The switch is turned on allowing current to begin flowing, then once the desired level is reached the switch is turned off and the current begins to fall. The ratio of ON time to OFF time (or Pulse Width) will determine the average current achieved.

The average current through the solenoid will vary with the work required by the solenoid (force and rate of movement), and this in turn drives the cost and size of the components in the drive circuitry. A power MOSFET is the preferred device for this kind of current modulation because of the devices low ON resistance and fast switching capability. In general, the size and cost of the power MOSFET will increase with the average power dissipated by the part. The average power in a MOSFET has two components, "Conduction Losses" and "Switching Losses." The conduction losses are directly proportional to the load current and ON resistance of the device. The switching losses are directly proportional to the system voltage, load current, the switching time of the device.

$$P_{Conduction} = I_{RMS}^2 \cdot R_{ON}$$

$$P_{Switching} = \frac{n}{T} \cdot \left(\frac{I_{ON} \cdot V_{DS}}{2} \cdot t_{Off} + \frac{I_{Off} \cdot V_{DS}}{2} \cdot t_{On} \right)$$

As can be seen from the preceding equations, for a given load current and system voltage, using a lower ON resistance part can reduce the conduction losses. This generally requires larger and more expensive devices for significant benefits. The other way to reduce power is by reducing switching times,

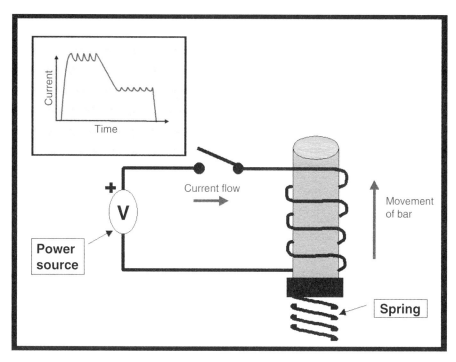

Figure 17.1   A simple solenoid

which may be achieved with a high performance "Gate Driver," and without the need to use newer, more expensive technology.

The power MOSFET can be placed in the "High-Side" (positive) or the "Low-Side" (negative) of the power source, or even a combination of both. Figure 17.1 illustrates a high-side configuration. The high side configuration is generally more challenging because of the need to provide a gate drive voltage, which is 10–15 V higher than the main system voltage. Several topologies are available to perform this function, but a "Bootstrap" or "Floating MOSFET Gate Driver" (Figure 17.2) is preferred because of cost. There are, however, several limitations of this circuit, particularly drive strength. Recognizing the popularity of this circuit topology, but also the shortcomings of it, several electronics suppliers offer integrated solutions (ICs) with very good performance. As should be expected, a premium is paid for the performance and convenience of the integrated solutions.

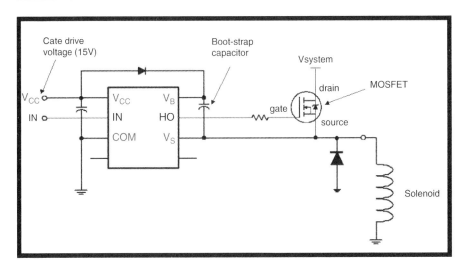

Figure 17.2   A floating MOS gate drive (IC)

17.3 Introduction

The main objective of this study was to develop a discrete floating MOS gate drive circuit with performance that was comparable to the existing IC solution. The major constraints were that an aggressive cost reduction must be achieved and minimal increase in circuit board real estate must be maintained. The current design used 24 Floating MOS Gate Driver ICs and so success would yield a very significant cost savings as well as a reusable "Building Block" that could be used in other designs.

An initial concept was developed using traditional methods; however, this design still exhibited very significant variation over the application operating range and consequently did not achieve the desired performance needed to replace the IC solution. At this point, the application of Taguchi Robust Design Methods was used to fine-tune the performance and reduce variability of the basic concept such that it could meet the application requirements. Figure 17.3 shows the baseline circuit topology.

17.4 Developing the "Ideal" Function

The ideal output response for this kind of circuit is to turn ON the MOSFET instantaneously. Although practically this is not achievable, this is the ideal response.

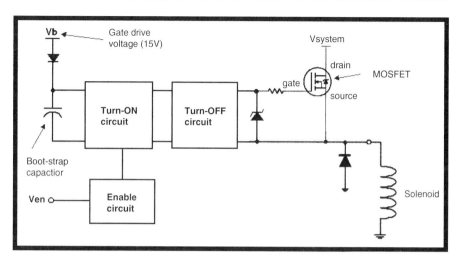

Figure 17.3 Baseline circuit topology

The MOSFET is turned on by charging the gate-to-source capacitance; an example waveform is shown in Figure 17.4.

So the ideal response of this circuit is a 0–1 Step function for turn-on, and 1-0 step function for turn-off. The obvious conclusion here would be to try to minimize the turn-on and turn-off times by using a "Smaller-the-Better" approach. The actual approach we took was to use a "Dynamic" function that described the ideal circuit response (Figure 17.5).

The ideal response was broken down into two parts; turn-on began from the instant the enable signal was asserted, $t_{ON}(0)$, and ended 10 us later; the turn-off began when the enable signal was negated, $t_{OFF}(0)$, and ended 10 us later. The 10 us period for each response was chosen so as to allow the output to stabilize. By using this approach, two dynamic functions were able to be formulated which completely described the "Ideal" response:

$$M = t - t_{ON}(0)$$

$$Y_{Turn-On}(M) = \int_{t_{ON}(0)}^{M} V_{gs}(M).dM$$

$$M = t - t_{OFF}(0)$$

$$Y_{Turn-Off}(M) = \int_{t_{OFF}(0)}^{M} [Vb - V_{gs}(M)].dM$$

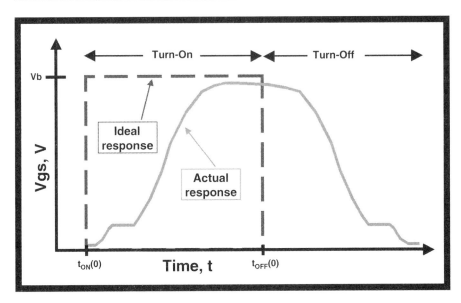

Figure 17.4   Ideal circuit response

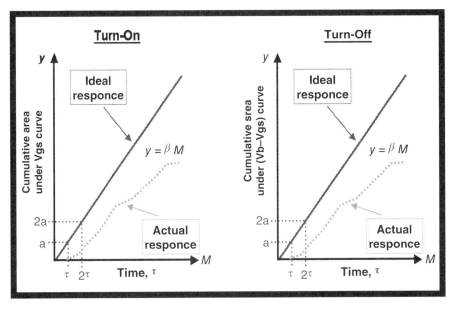

Figure 17.5   Dynamic ideal functions for Turn-On and Turn-Off

When applied to the ideal response these functions would yield a straight line, which is ideally suited for the dynamic signal-to-noise calculation.

There are several undesirable features of the practical response of this circuit:

- Delay-time from $t_{ON}(0)$ until the output begins to move from it's initial state.
- Low slew rate – this is the overall rate of change of the output, or the slope.
- Plateau length – this is the "flat-spot" that is characteristic of a MOSFET.
- Droop – applies to turn-on and is the amount the voltage drops from its maximum value within the event period (10 us).
- Saturation – this is the difference between Vb and the maximum Vgs for turn-on, and the difference between 0 (zero) and the minimum Vgs for turn-off.

By driving these ideal functions toward a straight line with a normalized slope of 1, all these undesirable features should be minimized as the response approaches a step response.

17.5 Noise Strategy

The predominant noise factors in this system are well known to be:

- Ambient operating temperature - −40 °C, 25 °C, 125 °C
- Transistor current gain tolerance - low, nominal, high.

These factors can be compounded into a single factor with three levels (Table 17.1).

17.6 Control Factors and Levels

The control factors in this experiment can be categorized as:

- key component nominal values;
- key component grade;
- subcircuit configurations.

Table 17.1 Single factor with three levels

	Ambient operating temperature	Transistor current gain tolerance
N1 – Nominal response	25 °C	Nominal
N2 – Slow response	−40 °C	Low
N3 – Fast response	125 °C	High

Table 17.2 Several control factor levels adjusted

	Level 1	Level 2	Level 3
Rpu_bst	High	Low	—
Transistor type	Low gain	Medium gain	High gain
Turn-On subcircuit configuration	Type 1	Type 2	Type 3
Turn-Off subcircuit configuration	Type 1	Type 2	Type 3
Rturn-on	High	Medium	Low
Rturn-off	High	Medium	Low
Rs_en	High	Medium	Low
Cen	Low	Medium	High

Before the final levels could be selected, some experimentation had to be done to test for "impossible combinations" or "boundary conditions." Much of this was learned during the initial concept development, however this exercise identified several boundary conditions that were unforeseen. Consequently, several control factor levels were adjusted (Table 17.2). The parameter diagram is shown in Figure 17.6.

17.7 Experiment Strategy and Measurement System

The experiment was performed using the Saber Simulation Environment. This method is more economical, more efficient and allows the designer to access parameter limits that would not be available using a prototype approach.

After parameter design simulation results were confirmed, a "bread-board" of the optimum and initial designs was constructed and tested with the worst-case conditions practically available (i.e., we cannot economically screen out the extremes in transistors).

These real samples were used primarily to confirm the validity of the SABER model.

17.8 Parameter Design Experiment Layout

The L18 orthogonal array was chosen for this experiment (Table 17.3). Each run comprised of the 3 levels of compounded noise and 1000 signal levels, in increments of 10 ns.

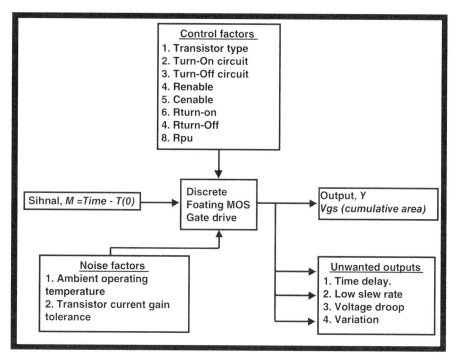

Figure 17.6 Parameter diagram

17.9 Results

See Figure 17.7(a) and Figure 17.7(b). From the raw data, it is clear that the real opportunity for improvement lies within the turn-on event. This holds true with observations made with the initial design and the nature of the two subcircuit topologies. It can be seen that there is significant variation over the experimental range, and that the optimum circuit yields a very significant improvement for the turn-on event. Although a significant improvement was not anticipated for the turn-off event, it was important to monitor it because strong interactions were observed during the development of the initial concept and control factors.

17.10 Response Charts

See Figures 17.8(a)–(d).

Table 17.3 L₁₈ orthogonal array

L₁₈ (2¹ × 3⁷)									M1			M2			⋯ ⋯	M1000			Ton		Toff	
No.	1	2	3	4	5	6	7	8	N1	N2	N3	N1	N2	N3	⋯ ⋯	N1	N2	N3	S/N	Beta	S/N	Beta
1	1	1	1	1	1	1	1	1											216.57	1.83	217.35	2.01
2	1	1	2	2	2	2	2	2											216.23	2.01	217.94	2.06
3	1	1	3	3	3	3	3	3											217.88	2.06	217.31	1.98
4	1	2	1	1	2	2	3	3											216.89	1.83	217.66	2.05
5	1	2	2	2	3	3	1	1											222.37	2.35	218.25	2.08
6	1	2	3	3	1	1	2	2							Since 3000 data points,				217.16	2.01	217.98	2.11
7	1	3	1	2	1	3	2	3							Row data are not shown				213.53	1.60	218.20	2.10
8	1	3	2	3	2	1	3	1											217.59	1.92	218.03	2.11
9	1	3	3	1	3	2	1	2											222.21	2.49	217.95	2.11
10	2	1	1	3	3	2	2	1											220.27	2.18	218.01	2.14
11	2	1	2	1	1	3	3	2											216.95	2.02	217.22	2.05
12	2	1	3	2	2	1	1	3											215.17	0.52	218.79	2.16
13	2	2	1	2	3	1	3	2											220.08	2.12	218.14	2.09
14	2	2	2	3	1	2	1	3											218.21	2.00	218.04	2.11
15	2	2	3	1	2	3	2	1											219.26	2.20	218.13	2.11
16	2	3	1	3	2	3	1	2											217.42	1.91	218.02	2.13
17	2	3	2	1	3	1	2	3											221.96	2.32	217.96	2.08
18	2	3	3	2	1	2	3	1											217.38	2.02	218.17	2.09

TURN-ON

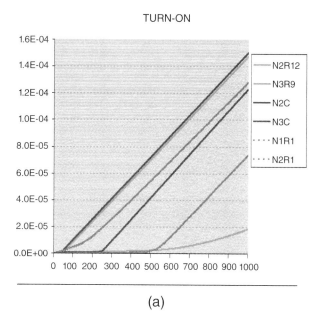

(a)

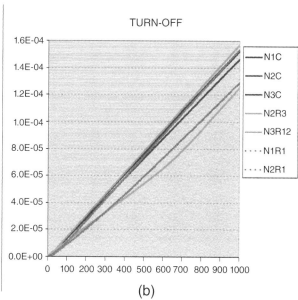

(b)

Figure 17.7   (a) Results for Turn-On event; (b) results for Turn-Off event

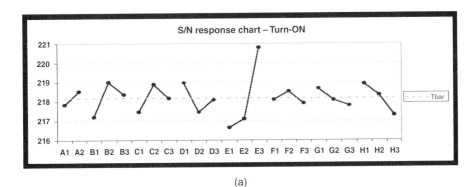

(a)

(b)

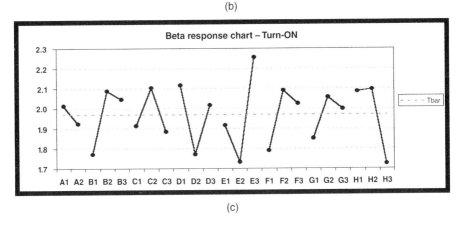

(c)

Figure 17.8   (a) S/N response chart – Turn-On; (b) S/N response chart – Turn-Off; (c) beta response graph – Turn-On; (d) beta response graph – Turn-Off

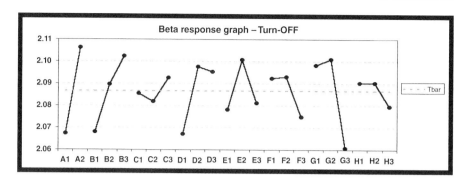

(d)

Figure 17.8 (*Continued*)

17.11 Two-Step Optimization

After studying the data, it was clear that there was an opportunity to significantly improve the turn-on event, and relatively little opportunity to improve the turn-off event. This data, combined with the fact that the turn-off event was already meeting the target specification, weighted the selection of the optimum factors in favor of the turn-on event. In general, the factor that yielded the largest S/N for turn-on was selected, unless that factor had a very negative effect on the beta of turn-on. Despite this bias toward the turn-on event, the optimum factors still presented some improvement for the turn-off event. See Table 17.4.

Table 17.4 Optimum factors presenting improvement for turn-off event

	A	B	C	D	E	F	G	H
Initial design	1	1	1	1	1	1	1	1
To maximize S/N for Ton	2	2	2	1	3	2	1	1
To maximize S/N for Toff	2	3	3	2	2	1	1	1
To maximize Beta for Ton	1	2	2	1	3	2	2	2
To maximize Beta for Toff	2	3	3	2	2	2	2	1
Optimum design	2	2	2	1	3	2	2	1

17.12 Confirmation

The predictions for the optimal design were as expected; a significant improvement of the S/N Ratio for turn-on was predicted and a relatively small improvement of the S/N Ratio for turn-off. The same trend was also true for the Beta. The optimum design confirmed very well with an actual gain of 6.141 dB for turn-on, which is approximately a 50% reduction in variability. The turn-off response, although relatively insignificant, still produced a gain of 1.57 dB, or a 16% reduction in variability. The same kind of improvement was confirmed for Beta, with a 41% increase in Beta for turn-on, and a 7% increase for turn-off. See Table 17.5(a) and Table 17.5(b).

The effects of Robust Optimization can be seen very clearly in the actual circuit response waveforms. The response is visibly "squarer" (more like the ideal step), the delay has been cut in half and the "plateau" region has been shortened by as much as 2/3. See the comparison of actual circuit responses shown in Figures 17.9(a) and 17.9(b).

17.13 Conclusions

Strong interactions between the turn-on and turn-off subcircuits were clearly identified. (optimum turn-on circuit may actually cause problems with the turn-off circuit, and vice-versa). Most of the circuit parameters have a significant effect on the turn-on response, but have a relatively insignificant effect on the turn-off response. Contrary to initial assumptions, it was found

Table 17.5 Prediction and confirmation: (a) Turn-On; (b) Turn-Off

Turn-ON	Prediction		Confirmation	
	S/N Ratio	Beta	S/N Ratio	Beta
Initial Design	216.511	1.681	216.574	1.833
Optimum Design	224.459	2.939	222.715	2.587
Gain	7.948	1.258	6.141	0.754

(a)

Turn-OFF	Prediction		Confirmation	
	S/N Ratio	Beta	S/N Ratio	Beta
Initial Design	217.484	2.040	217.346	2.010
Optimum Design	217.964	2.103	218.916	2.156
Gain	0.480	0.063	1.569	0.146

(b)

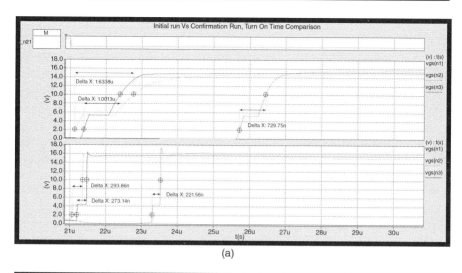

(a)

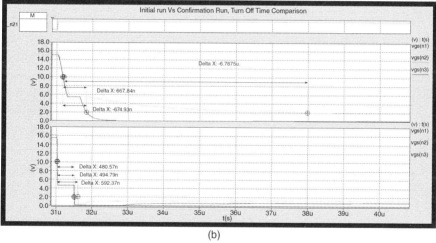

(b)

Figure 17.9   Comparison of actual circuit response: (a) Turn-On; (b) Turn-Off

that the highest gain transistors (most expensive) were not necessarily better, due to turn-on/off circuit interactions.

By using Robust Design Methods, an optimum circuit configuration was selected, which provided the fastest turn-on/off time whilst minimizing variation with temperature and component variation. A significant performance improvement was achieved over the baseline design. The Turn-On response variation was reduced by 50% (Mostly Delay) and the Turn-On time was reduced by 70–80% of the initial circuit. The Turn-Off response variation was reduced by 16% (Mostly Saturation Voltage) and the Turn-Off time was reduced by 20% of initial circuit. The overall circuit response has been improved by eliminating all categories of unwanted outputs. The responses are much "cleaner" and exhibit less saturation voltage limitation, time delay, slew-rate limiting and variation.

The optimum design was able to meet the required specification at a cost of 1/5 of the cost of the existing IC solution. The total time taken for this experiment was about 1 week, 3 days of which was to run the simulations and format the data. This experiment yielded a design with much improved performance over the initial "best effort" design in a similar time frame. Robust Engineering techniques optimized the design cycle time by identifying circuit interactions that are not necessarily intuitive.

The selection of a good ideal function was most critical in this experiment. By driving the circuit response toward a "true ideal." all elements of unwanted output can be reduced simultaneously. This is by far more efficient than focusing on individual symptoms.

17.13.1 Acknowledgments

Thanks to Brad Walker of ASI Consulting Group, LLC and Fred Kuhlman of Delphi Automotive Systems for their guidance as coaches on this project.

Thanks to Alan Wu of ASI Consulting Group, LLC for pulling the Ideal Function out of thin air.

This case study is contributed by Thomas Hedges and John Schieffer of Delphi-Delco Electronic Systems, USA.

18

Reformer Washcoat Adhesion on Metallic Substrates

Delphi Automotive Systems, USA

18.1 Executive Summary

It is the goal of this study to determine the control factors which govern adhesion of a reformer washcoat to a reformer substrate. The reformer catalyst is a noble metal, from the platinum-group metals that includes platinum, rhodium, palladium, etc. As platinum-group metals are very expensive, it is not feasible to increase the amount of precious metal used to compensate for catalyst loss during reformer operation. Hence, the amount of washcoat applied must remain adhered to the substrate throughout the product lifetime. Catalyst loss is a problem for reformers due to spallation of the washcoat. Taguchi Robust Engineering methods were chosen to investigate the design space for reformer washcoat adhesion using an L54 matrix with 3 duplicates for each design factor combination or run. After implementing the results from the study, the reformer washcoat has successfully maintained its washcoat for 480 hours and 30 thermal cycles without excessive loss of catalyst.

Robust Optimization: World's Best Practices for Developing Winning Vehicles,
First Edition. Subir Chowdhury and Shin Taguchi.
© 2016 Subir Chowdhury, Shin Taguchi, and ASI Consulting Group, LLC.
Published 2016 by John Wiley & Sons, Ltd.

18.2 Introduction

Catalytic reformers convert liquid and gaseous fuels into a hydrocarbon gas, termed "reformate." A catalytic reformer operates between 850 and 1100 °C and must be capable of withstanding many thermal cycles over tens of thousands of hours of operation. Figure 18.1 is a schematic showing two different geometries of a reformer.

Fuel and air are introduced into the reformer where it is reformed. A washcoat reforms the fuel and air into reformate. The washcoat is a slurry mixture of metal catalyst and a ceramic support. The reformer substrates, which hold the washcoat, can have several different geometries as shown in Figure 18.2.

Both ceramic and metal materials can be used as reformer substrates. However, ceramic foam substrates are brittle, fragile materials which restrict reformer designs due to the lack of robustness of the material. Further advancement in reformer design requires the use of materials which allow greater versatility in design without sacrificing robustness. Since metallic materials have greater ductility and toughness versus their ceramic counterparts, metallic substrates are more attractive for these advanced reformer

Figure 18.1 Concepts of a tubular reformer and a planar reformer

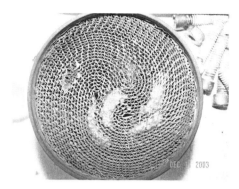

Figure 18.2 Various possible geometries for reformer substrates

designs. The purpose of this study is to investigate various, metallic materials as candidates for a reformer substrate. During operation, the washcoat must stay adhered to the substrate. The loss of washcoat and, therefore, catalyst in the reformer leads to decreased reformer efficiency and increased higher hydrocarbon formation which is detrimental to reformer performance.

Metallic substrates differ appreciably from ceramic substrates due to the differences in the oxidation rates for metal and ceramic substrates between 800 and 1100 °C. Ceramic materials, such as zirconia toughened alumina, ZTA, can be considered as static in this temperature range. Metallic materials in this temperature range, however, are dynamic since oxidation of the metal causes the interface between the washcoat and the metal to change with time. The metal will oxidize unless a protective coating is applied to the surface. If the oxide scale, which forms on the metal substrate spalls, then the catalyst will also spall. Hence, it is vital for the oxide scale to remain well-adhered to the substrate during operation. The washcoat also must remain attached to any oxide scales which form. Adherence can be particularly difficult under thermal cycling conditions. It is important that the oxide scale does not grow so fast that it engulfs the washcoat rendering the catalyst useless during operation.

18.3 Experimental Setup

Taguchi robust engineering methods were used to develop an L54 orthogonal matrix to study the effect of metallic substrate composition and process controls on washcoat adhesion. Because so many choices were needed for

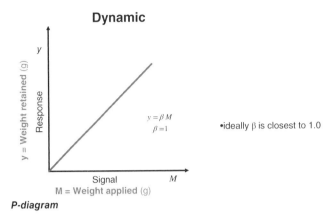

Figure 18.3   The ideal function for the L54 washcoat adhesion study

the alloy composition control factor, columns 2 and 9–14 were combined to form the 18-level, column 1. Figure 18.3 shows the blank L54 orthogonal matrix used in this study.

18.3.1   The Ideal Function

The ideal function is graphed in Figure 18.3. The signal, M for the ideal function is the weight of washcoat applied to the coupon in units of grams. The response, y of the ideal function is the weight of washcoat retained on the coupon in units of grams. The ideal function is then represented by the equation for a straight line $y = \beta M$. The slope of the straight line is beta, β which will ideally have the target value of 1.0, as shown in Figure 18.3. Data, which falls below the $\beta = 1.0$-line, indicates that weight loss has occurred; while data, which falls above the line, indicates weight gain. The further from the $\beta = 1.0$-line, the more extreme the weight gain or loss and, consequently, the further from ideal behavior.

18.3.2   P-Diagram

The parameter diagram, P-diagram shown in Figure 18.4 is a schematic of the inputs to the washcoat adhesion system and relates them to the outputs. The P-diagram includes the control factors, noise factors, signal factor (input), response factor (output) and unwanted outputs. The factors examined in this study are listed in Figure 18.4.

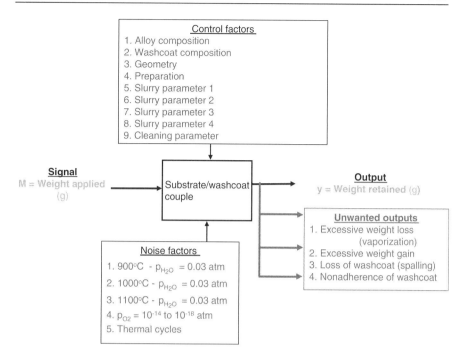

Figure 18.4   The P-diagram for the L54 washcoat adhesion study

18.3.3   Control Factors

18.3.3.1   Alloy Composition

This study investigated 18 different alloy compositions including nickel-based superalloys, austenitic stainless steels, and ferritic stainless steels, and Fe-based superalloys. These alloys consisted of both chromia-formers and alumina-formers.

Silica-formers were not considered for this application. SiO_2 is known to decompose in a high temperature, low partial pressure of oxygen atmosphere into gaseous forms of Si, such as SiO_3, etc. Stainless steel represents the ideal material for a reformer substrate in terms of cost; however, the oxidation performance of stainless steel makes the use of these materials difficult. Nickel-based alloys have the oxidation resistance at operating temperature. Both alumina- and chromia-forming alloys were in this study. Some of the compositions were coatings which were applied to the base alloy. Coatings were not added as a separate control factor because several of the alloy/coating couples were incompatible and several missing data approximations would have been necessary. Instead the compatible alloy/coating couples were treated as a separate choice in the composition column.

18.3.3.2 Washcoat Composition

There were three different washcoat compositions chosen. Each washcoat factor level consisted of the same noble metal catalyst component and a ceramic slurry matrix component. There was a different ceramic slurry matrix component for each level of this control factor.

18.3.3.3 Slurry Parameters

There were four different slurry parameters chosen as control factors.

18.3.3.4 Cleaning Procedures

The cleaning procedures chosen for the coupons varied dramatically.

18.3.3.5 Preparation

Three different coupon preparation were used as a control factor.

18.4 Control Factor Levels

The levels for each factor are described in Table 18.1.

18.5 Noise Factors

The noise factors are variables that affect a system but are not controllable or are too expensive to control. The noise factors in this study were temperature and thermal cycling. These noise factors were chosen because the reformer needs to operate at a variety of temperatures and the number of thermal cycles will be variable depending on the application.

18.5.1 Signal Factor

The signal factor for this study is weight change (g).

18.5.2 Unwanted Outputs

Unwanted outputs are the conditions spurious conditions which need to be avoided in the experiment. Unwanted outputs for the study are excessive

Table 18.1 Control factors and levels for L54 washcoat adhesion study

Level	Alloy composition	Washcoat composition	Geometry	Preparation	Slurry parameter 1	Slurry parameter 2	Slurry parameter 3	Slurry parameter 4	Cleaning
1	A1/coat 3	WC1	C1	P1	E1	F1	G1	H1	J1
2	A5	WC2	C2	P2	E2	F2	G2	H2	J2
3	A7	WC3	C3	P3	E3	F3	G3	H3	J3
4	A2								
5	A6								
6	A4								
7	A3								
8	A5/coat 1								
9	A7/coat 1								
10	A4/coat 1								
11	A2/coat 3								
12	A8								
13	A3/coat 3								
14	A5/coat 2								
15	A7/coat 2								
16	A4/coat 2								
17	A9								
18	A1								

weight loss due to vaporization or spalling. Excessive weight gain due to oxidation is an indication of catastrophic oxidation. Spalling is the spontaneous separation of a coating from a surface. It is possible that a coupon could first spall its washcoat and then oxidize such that the overall weight change is unchanged. This is accounted for by oxidizing the coupons in individual separate porcelain crucibles. Any spalled material was captured and easily observed in the crucibles. Any time spalled material was observed in a crucible, all the samples were weighed and the data was recorded. The spalled material was then discarded and no longer weighed. During testing catastrophic oxidation was observed in the austenitic and ferritic stainless steels at 1000 and 1100 °C. A missing data approximation had to be done for these alloy compositions at these temperatures. The experiment was operated below the vaporization temperature of the metal catalysts to prevent the unwanted output.

18.6 Description of Experiment

A reformer substrate operates between 800 and 1100 °C and the substrate must be as robust at the first thermal cycle and the last thermal cycle.

18.6.1 Furnace

The furnace is a CM Model 1212 FL with Retort. The testing atmosphere was a hydrogen-water-nitrogen mixture. Equation (18.1) is the calculation for water vapor in a gas

$$\frac{-2961}{T_{Bubbler}} - 5.13\,LOG(T_{Bubbler}) + 21.133 = LOG(p_{H_2O}) \qquad (18.1)$$

Equation (18.2) is the calculation for the Gibb's free energy for hydrogen and oxygen converting to water vapor.

$$H_2 + 1/2\,O_2 = H_2O$$
$$\Delta G = \frac{-1,100,000\,T + 50,000\,T^2 + 510,000}{RT} \qquad (18.2)$$

Equation (18.3) is the calculation for the partial pressure of oxygen, p_{O_2} for the conversion of hydrogen and oxygen into water vapor.

$$p_{O_2} = \frac{(p_{H_2O})^2}{p_{H_2}^2}\,\exp\left[\frac{-246,000 + 54.8\,T}{8.3144\,T}\right]^2 \qquad (18.3)$$

Coupons were tested at 900, 1000, and 1100 °C. The calculation of the water vapor in the testing atmosphere gave varied p_{O_2}'s of 2×10^{-17} atm at 900 °C, 8×10^{-16} atm at 1000 °C, and 2×10^{-14} atm at 1100 °C. There were 3 replicates for each design combination.

The coupons were thermally cycled every 16 hours. The coupons tested at 1100 °C, showed the severest oxidation and those at 900 °C showed the least. The coupons were tested for 6 thermal cycles with a total oxidation time of 96 hours. The testing was then stopped due to excessive oxidation of most of the coupons at 1100 °C.

18.6.2 Orthogonal Array and Inner Array

Table 18.2 is the orthogonal array for this study. It shows the parameters for each of the 54 design combinations. Table 18.3 is the inner array for the experimental testing. It shows the strategy for the running the experiments. Three sets of 54 combinations with 3 replicates for each design combination were made and oxidized at 900, 1000 or 1100 °C.

18.6.3 Signal-to-Noise and Beta Calculations

The signal-to-noise ratio and beta calculations for this study are shown in Table 18.4.

The equations for these calculations can be found in the *Robust Engineering Handbook* [2].

18.6.4 Response Tables

Tables 18.5 and 18.6 are the response tables for the signal-to-noise ratio and beta. The ranking in the tables clearly demonstrates that the substrate composition is the dominant factor for all the control factors investigated. The preparation of the coupon was the distant second.

Figure 18.5 is a diagram of the evaluation guidelines for determining the best choices for signal-to-noise, S/N and beta in the response graphs in Figure 18.6 and Figure 18.7.

18.7 Two Step Optimization and Prediction

Since the alloy composition was such an important control factor, four alloy choices were chosen for confirmation.

Table 18.2 Outer orthogonal array for washcoat adhesion study

L54	Substrate material	Washcoat comp	Geometry	Prep	Slurry param 1	Slurry param 2	Slurry param 3	Slurry param 4	Clean
	1*	3	4	5	6	7	8	15	16
1	A1/coat 3	WC1	C1	P1	E1	F1	G1	H1	J1
2	A5	WC1	C1	P1	E1	F1	G1	H2	J2
3	A7	WC1	C1	P1	E1	F1	G1	H3	J3
4	A1/coat 3	WC2	C2	P2	E2	F2	G2	H2	J3
5	A5	WC2	C2	P2	E2	F2	G2	H3	J1
6	A7	WC2	C2	P2	E2	F2	G2	H1	J2
7	A1/coat 3	WC3	C3	P3	E3	F3	G3	H3	J2
8	A5	WC3	C3	P3	E3	F3	G3	H1	J3
9	A7	WC3	C3	P3	E3	F3	G3	H2	J1
10	A2	WC1	C1	P2	E2	F3	G3	H1	J1
11	A6	WC1	C1	P2	E2	F3	G3	H2	J2
12	A4	WC1	C1	P2	E2	F3	G3	H3	J3
13	A2	WC2	C2	P3	E3	F1	G1	H2	J3
14	A6	WC2	C2	P3	E3	F1	G1	H3	J1
15	A4	WC2	C2	P3	E3	F1	G1	H1	J2
16	A2	WC3	C3	P1	E1	F2	G2	H3	J2
17	A6	WC3	C3	P1	E1	F2	G2	H1	J3
18	A4	WC3	C3	P1	E1	F2	G2	H2	J1
19	A3	WC1	C2	P1	E3	F2	G3	H1	J1
20	A5/coat 1	WC1	C2	P1	E3	F2	G3	H2	J2
21	A7/coat 1	WC1	C2	P1	E3	F2	G3	H3	J3
22	A3	WC2	C3	P2	E1	F3	G1	H2	J3
23	A5/coat 1	WC2	C3	P2	E1	F3	G1	H3	J1
24	A7/coat 1	WC2	C3	P2	E1	F3	G1	H1	J2
25	A3	WC3	C1	P3	E2	F1	G2	H3	J2
26	A5/coat 1	WC3	C1	P3	E2	F1	G2	H1	J3
27	A7/coat 1	WC3	C1	P3	E2	F1	G2	H2	J1
28	A4/coat 1	WC1	C3	P3	E2	F2	G1	H1	J1
29	A2/coat 3	WC1	C3	P3	E2	F2	G1	H2	J2
30	A8	WC1	C3	P3	E2	F2	G1	H3	J3
31	A4/coat 1	WC2	C1	P1	E3	F3	G2	H2	J3
32	A2/coat 3	WC2	C1	P1	E3	F3	G2	H3	J1
33	A8	WC2	C1	P1	E3	F3	G2	H1	J2
34	A4/coat 1	WC3	C2	P2	E1	F1	G3	H3	J2
35	A2/coat 3	WC3	C2	P2	E1	F1	G3	H1	J3
36	A8	WC3	C2	P2	E1	F1	G3	H2	J1
37	A3/coat 3	WC1	C2	P3	E1	F3	G2	H1	J1
38	A5/coat 2	WC1	C2	P3	E1	F3	G2	H2	J2
39	A7/coat 2	WC1	C2	P3	E1	F3	G2	H3	J3
40	A3/coat 3	WC2	C3	P1	E2	F1	G3	H2	J3
41	A5/coat 2	WC2	C3	P1	E2	F1	G3	H3	J1
42	A7/coat 2	WC2	C3	P1	E2	F1	G3	H1	J2
43	A3/coat 3	WC3	C1	P2	E3	F2	G1	H3	J2
44	A5/coat 2	WC3	C1	P2	E3	F2	G1	H1	J3
45	A7/coat 2	WC3	C1	P2	E3	F2	G1	H2	J1
46	A4/coat 2	WC1	C3	P2	E3	F1	G2	H1	J1
47	A9	WC1	C3	P2	E3	F1	G2	H2	J2
48	A1	WC1	C3	P2	E3	F1	G2	H3	J3
49	A4/coat 2	WC2	C1	P3	E1	F2	G3	H2	J3
50	A9	WC2	C1	P3	E1	F2	G3	H3	J1
51	A1	WC2	C1	P3	E1	F2	G3	H1	J2
52	A4/coat 2	WC3	C2	P1	E2	F3	G1	H3	J2
53	A9	WC3	C2	P1	E2	F3	G1	H1	J3
54	A1	WC3	C2	P1	E2	F3	G1	H2	J1

Table 18.3 Inner Array for L54 washcoat adhesion study

	Weights after coating								
	M1			M2			M3		
L54	900	1000	1100	900	1000	1100	900	1000	1100
1	2.304	2.431	2.456	2.455	2.300	2.377	2.441	2.453	2.296
2	1.035	1.025	1.028	1.038	1.013	1.040	1.014	1.014	1.021
3	1.769	1.760	1.762	1.763	1.770	1.754	1.748	1.772	1.771
4	1.399	1.392	0.968	1.490	1.261	1.352	1.199	1.304	1.358
5	1.054	1.085	1.076	1.068	1.057	1.065	1.044	1.075	1.088
6	1.704	1.763	1.724	1.592	1.679	1.702	1.614	1.658	1.723
7	1.541	1.530	1.473	1.483	1.534	1.504	1.514	1.532	1.364
8	1.101	1.058	1.057	1.058	1.050	1.044	1.057	1.064	1.073
9	1.850	1.899	1.870	1.837	1.920	1.860	1.874	1.857	1.854
10	0.637	0.684	0.693	0.607	0.602	0.706	0.685	0.686	0.628
11	21.067	20.751	20.950	20.725	21.102	20.725	20.942	21.065	21.058
12	1.415	1.391	1.415	1.726	1.725	1.527	1.690	1.565	1.619
13	0.839	0.815	0.833	0.832	0.842	0.838	0.830	0.818	0.834
14	21.242	20.969	21.232	21.374	21.304	21.365	20.933	20.983	20.855
15	1.999	1.967	1.988	1.961	1.998	1.989	1.990	1.965	2.012
16	0.771	0.765	0.777	0.770	0.771	0.768	0.777	0.773	0.775
17	20.842	21.041	21.234	20.899	21.193	21.008	21.341	21.074	21.291
18	1.887	1.870	1.907	1.890	1.871	1.851	1.912	1.913	1.916
19	3.363	3.368	3.388	3.404	3.370	3.373	3.371	3.354	3.380
20	1.376	1.313	1.353	1.147	1.172	1.161	1.176	1.175	1.188
21	2.000	2.114	2.109	1.912	1.932	1.885	1.864	1.875	1.899
22	3.447	3.446	3.459	3.446	3.468	3.451	3.432	3.450	3.430
23	1.288	1.545	1.568	1.436	1.460	1.533	1.539	1.034	1.565
24	1.997	1.956	2.229	2.032	2.097	1.853	2.208	2.035	2.081
25	3.397	3.420	3.434	3.409	3.406	3.434	3.391	3.402	3.399
26	1.318	1.252	1.328	1.159	1.159	1.148	1.147	1.204	1.144
27	2.783	2.757	2.740	2.588	2.541	2.482	2.271	2.434	2.469
28	2.409	2.311	2.482	2.401	2.312	2.275	2.406	2.348	2.415
29	0.786	1.614	lost	1.225	0.742	1.270	1.322	1.305	1.110
30	22.278	21.724	21.843	22.450	22.736	22.966	22.536	22.803	22.623
31	2.395	2.500	2.428	2.410	2.491	2.410	2.361	2.572	2.359
32	0.981	0.746	1.117	0.917	0.900	0.924	0.899	0.887	0.878
33	0.974	0.964	0.972	0.969	0.997	0.979	0.987	0.964	0.970
34	2.139	2.243	1.819	1.950	1.837	1.840	1.883	1.826	1.946
35	0.615	0.757	1.273	1.121	0.758	1.032	1.350	0.808	1.062
36	21.290	0.815	21.271	22.879	0.812	21.945	22.372	0.771	21.923
37	3.828	3.785	3.557	3.747	3.818	3.893	3.845	3.792	3.616
38	1.968	1.804	1.888	1.986	1.792	1.781	1.611	1.579	1.708
39	2.096	2.077	2.057	1.876	1.871	1.834	1.949	1.914	1.897
40	3.860	3.832	3.891	3.614	3.436	3.543	3.565	3.506	3.496
41	1.405	1.338	1.503	1.276	1.278	1.342	1.253	1.285	1.312
42	2.362	2.345	2.235	2.419	2.382	2.422	2.277	2.471	2.456
43	3.617	3.590	3.571	3.545	3.435	3.433	3.544	3.514	3.448
44	1.975	1.833	1.974	1.755	1.839	1.754	1.722	1.656	1.642
45	2.167	2.007	2.095	2.569	2.568	2.304	2.340	2.347	2.368
46	2.245	2.531	2.452	2.509	2.622	2.135	2.548	2.415	2.180
47	0.830	0.855	0.846	0.832	0.819	0.883	0.855	0.816	0.842
48	0.815	0.928	0.856	0.869	0.810	0.875	0.857	0.931	0.859
49	2.144	2.211	2.108	1.915	1.950	1.957	1.989	1.966	1.976
50	0.575	0.580	0.578	0.575	0.575	0.576	0.580	0.579	0.580
51	1.068	1.079	1.074	1.075	1.043	1.056	1.069	1.075	1.062
52	2.688	2.669	2.724	2.720	2.737	2.719	2.565	2.775	2.564
53	0.594	0.592	0.583	0.593	0.594	0.590	0.599	0.595	0.596
54	1.111	1.113	1.098	1.115	1.106	1.114	1.119	1.115	1.112

Table 18.4 Signal-to-noise and beta calculations for L54 washcoat

L54	Substrate material	r	r_0	S_T	S	S_{XN}	S_e	V_e	V_N	$S - V_e$	S/N	Slope
1	A1/coat 3	0.001	3	0.141	0.020	0.009	0.112	0.005	0.005	0.016	29.19	1.96
2	A5	0.007	3	0.014	0.013	-0.006	0.008	0.000	0.000	0.013	40.97	0.78
3	A7	0.000	3	0.002	0.001	-0.001	0.001	0.000	0.000	0.001	47.12	1.27
4	A1/coat 3	0.059	3	0.849	0.480	-0.201	0.569	0.024	0.014	0.457	22.63	1.61
5	A5	0.065	3	0.184	0.183	-0.112	0.113	0.005	0.000	0.178	43.74	0.96
6	A7	0.083	3	0.199	0.162	-0.092	0.129	0.005	0.001	0.156	26.43	0.79
7	A1/coat 3	0.084	3	0.364	0.275	-0.178	0.267	0.011	0.003	0.264	24.87	1.03
8	A5	0.044	3	0.132	0.130	-0.079	0.081	0.003	0.000	0.127	41.21	0.98
9	A7	0.099	3	0.194	0.185	-0.120	0.129	0.005	0.000	0.180	32.27	0.78
10	A2	0.001	3	0.047	0.022	-0.007	0.032	0.001	0.001	0.020	37.04	2.24
11	A6	0.020	3	0.360	0.100	-0.047	0.307	0.013	0.010	0.087	21.60	1.20
12	A4	0.009	3	0.715	0.227	-0.036	0.523	0.022	0.019	0.206	25.90	2.70
13	A2	0.068	3	0.169	0.166	-0.110	0.112	0.005	0.000	0.162	39.77	0.89
14	A6	0.093	3	0.630	0.396	-0.215	0.449	0.019	0.009	0.378	21.79	1.17
15	A4	0.090	3	0.073	0.050	-0.028	0.052	0.002	0.001	0.047	22.81	0.42
16	A2	0.008	3	0.032	0.031	-0.020	0.021	0.001	0.000	0.030	44.45	1.09
17	A6	0.019	3	0.369	0.171	-0.034	0.232	0.010	0.008	0.162	25.63	1.67
18	A4	0.002	3	0.020	0.013	-0.006	0.014	0.001	0.000	0.012	38.71	1.48
19	A3	0.003	3	0.033	0.020	-0.011	0.025	0.001	0.001	0.019	36.90	1.58
20	A5/coat 1	0.035	3	0.284	0.186	-0.108	0.206	0.009	0.004	0.178	26.57	1.31
21	A7/coat 1	0.031	3	0.243	0.091	0.042	0.110	0.005	0.006	0.087	22.05	0.97
22	A3	0.108	3	0.382	0.373	-0.246	0.255	0.011	0.000	0.363	35.25	1.06
23	A5/coat 1	0.116	3	1.165	0.512	-0.245	0.898	0.037	0.025	0.474	17.35	1.17
24	A7/coat 1	0.041	3	0.616	0.301	-0.096	0.411	0.017	0.012	0.284	22.81	1.52
25	A3	0.022	3	0.096	0.092	-0.058	0.062	0.003	0.000	0.089	39.44	1.17
26	A5/coat 1	0.017	3	0.145	0.088	-0.040	0.097	0.004	0.002	0.084	28.67	1.27
27	A7/coat 1	0.051	3	0.103	0.058	-0.034	0.078	0.003	0.002	0.055	23.24	0.60
28	A4/coat 1	0.094	3	0.480	0.361	-0.195	0.313	0.013	0.005	0.348	24.33	1.11
29	A2/coat 3	0.122	3	1.017	0.635	-0.118	0.500	0.021	0.015	0.615	20.60	1.30
30	A8	0.146	3	4.117	1.434	-0.644	3.326	0.139	0.103	1.296	14.58	1.72
31	A4/coat 1	0.056	3	0.668	0.448	-0.167	0.387	0.016	0.008	0.432	24.79	1.60
32	A2/coat 3	0.051	3	0.501	0.251	-0.130	0.379	0.016	0.010	0.236	22.08	1.25
33	A8	0.099	3	0.559	0.484	-0.296	0.372	0.015	0.003	0.468	27.35	1.26
34	A4/coat 1	0.450	3	1.875	0.921	-0.314	1.268	0.053	0.037	0.868	12.44	0.80
35	A2/coat 3	0.012	3	1.540	0.565	-0.203	1.178	0.049	0.037	0.516	25.77	3.76
36	A8	0.017	3	6.667	1.051	0.187	5.428	0.226	0.216	0.825	18.64	3.98
37	A3/coat 3	0.056	3	0.657	0.333	-0.133	0.457	0.019	0.012	0.314	21.76	1.37
38	A5/coat 2	0.056	3	0.281	0.180	-0.100	0.201	0.008	0.004	0.171	24.19	1.01
39	A7/coat 2	0.003	3	0.002	0.000	0.000	0.002	0.000	0.000	0.000	21.58	0.10
40	A3/coat 3	0.024	3	0.128	0.078	-0.030	0.080	0.003	0.002	0.074	27.30	1.01
41	A5/coat 2	0.271	3	0.511	0.334	-0.167	0.344	0.014	0.007	0.320	17.63	0.63
42	A7/coat 2	0.212	3	0.811	0.736	-0.475	0.550	0.023	0.003	0.713	25.88	1.06
43	A3/coat 3	0.016	3	0.133	0.098	-0.056	0.091	0.004	0.001	0.094	31.59	1.40
44	A5/coat 2	0.162	3	0.665	0.522	-0.277	0.420	0.018	0.006	0.504	22.75	1.02
45	A7/coat 2	0.085	3	0.648	0.406	-0.109	0.350	0.015	0.009	0.391	22.20	1.24
46	A4/coat 2	0.086	3	1.316	0.694	-0.303	0.925	0.039	0.024	0.656	20.26	1.59
47	A9	0.001	3	0.052	0.021	-0.010	0.041	0.002	0.001	0.019	36.34	2.26
48	A1	0.000	3	0.392	0.023	-0.001	0.370	0.015	0.014	0.007	7.00	0.04
49	A4/coat 2	0.001	3	0.066	0.013	0.000	0.052	0.002	0.002	0.011	35.42	2.65
50	A9	0.002	3	0.028	0.019	-0.011	0.020	0.001	0.000	0.019	39.30	1.72
51	A1	0.023	3	0.200	0.112	-0.070	0.158	0.007	0.003	0.105	26.50	1.23
52	A4/coat 2	0.047	3	0.422	0.249	-0.105	0.279	0.012	0.007	0.237	24.04	1.30
53	A9	0.008	3	0.033	0.030	-0.019	0.022	0.001	0.000	0.029	40.55	1.12
54	A1	0.086	3	0.354	0.317	-0.196	0.234	0.010	0.001	0.307	29.13	1.09

Table 18.5 Response table for the signal-to-noise ratio

S/N response table

Level	Substrate material	Washcoat composition	Geometry	Prepar-ation	Slurry parameter 1	Slurry 2	Slurry parameter 3	Slurry parameter 4	Cleaning
1	25.56271	26.55509613	30.284841	30.57416	29.281772	26.9038189	28.1561166	28.10206	27.53122
2	41.97138	27.71219279	26.71198	24.98575	27.3732918	29.1312593	27.6832318	28.8669078	27.71553
3	35.273643	29.19976208	26.470227	27.90714	26.8119872	27.4319728	27.6277026	26.49808	28.2203
4	40.421558								
5	23.007829								
6	29.137852								
7	37.195748								
8	24.194743								
9	22.699827								
10	20.51968								
11	22.817633								
12	20.192738								
13	26.8844								
14	21.522548								
15	23.219382								
16	26.575755								
17	38.730261								
18	20.874619								
Δ	21.778642	2.644665953	3.8146143	5.588406	2.46978477	2.22744042	0.52841398	2.3688288	0.689071
Ranking	1	4	3	2	5	7	9	6	8

Evaluation Guidelines

Maximize S/N and Beta closest to 1.0

High robustness to temp range, cycles, Beta close to 1.0 **Consistently good** Low degree of weight change	High robustness to temp range, cycles, Beta close to 1.0 **Consistently bad** High degree of weight change
Low robustness to temp range, cycles **Inconsistently good** Low degree of weight change	Low robustness to temp range, cycles, **Inconsistently bad** High degree of weight change

S/N ratio

Figure 18.5   Evaluation guidelines for analysis

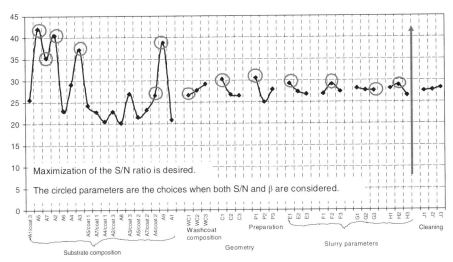

Figure 18.6   Graph of the signal-to-noise ratio illustrating importance of substrate composition to robustness of washcoat adhesion study

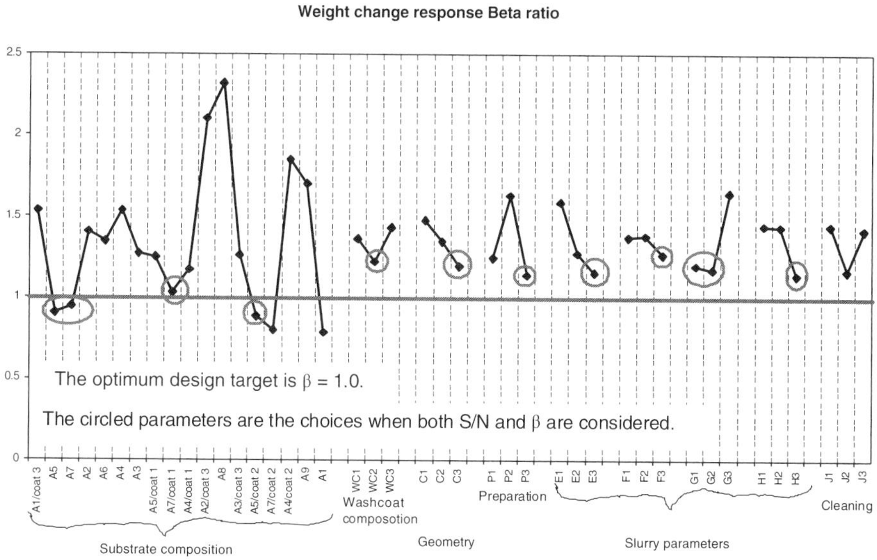

Figure 18.7 Graph of weight change response for beta ratio.

18.7.1 Optimum Design

Table 18.7.

18.7.2 Predictions

Four predictions were made for this study due to the size of the study. Table 18.8.

18.8 Confirmation

Tables 18.9(a)–(d).

A robust study is successful if the percentage difference in the gain is less than 33%.

Alloy compositions A5, A3 and A2 confirmed. Alloy composition A7 did far better than predicted. This was investigated further to find an interaction between the substrate composition and the washcoat composition.

18.8.1 Design Improvement

Figure 18.8 is the original design combination for the reformer prior to the robust engineering study. The graph shows that at 900 °C the original

Table 18.6 Response table for the beta parameter

Beta table

Level	Substrate material	Washcoat composition	Geometry	Prepar-ation	Slurry parameter 1	Slurry parameter 2	Slurry parameter 3	Slurry parameter 4	Cleaning
1	1.5337759	1.361762156	1.4747172	1.24484	1.58951638	1.37046849	1.19620903	1.44161	1.438664
2	0.903438	1.221317134	1.34579	1.630242	1.27122366	1.37960342	1.17290872	1.43603269	1.162431
3	0.9464416	1.431488613	1.194056	1.139485	1.15382787	1.264496	1.64545016	1.13692	1.413473
4	1.4049108								
5	1.3450902								
6	1.5335265								
7	1.2691984								
8	1.2476136								
9	1.0298149								
10	1.1705303								
11	2.102637								
12	2.317621								
13	1.2596582								
14	0.8863103								
15	0.8006307								
16	1.8481698								
17	1.7012256								
18	0.7868145								
Δ	1.530807	0.2102	0.2807	0.4908	0.4357	0.1151	0.4725	0.3047	0.2762
Ranking	1	8	6	2	4	9	3	5	7

Table 18.7 Selection of optimum design configurations

Substrate material/ thermal	Washcoat composition	Substrate geometry	Substrate surface finish	Binder (aluminum hydroxide)	Particle size (microns)	Washcot pH	Slurry aging time	Substrate detergent rinse
A7	WC3	C1	P3	E1	F2	G1	H2	J2
A5	WC3	C1	P3	E1	F2	G1	H2	J2
A3	WC3	C1	P3	E1	F2	G1	H2	J2
A2	WC3	C1	P3	E1	F2	G1	H2	J2

Table 18.8 Prediction of S/N and β for optimum configurations

	Predictions	
Substrate	S/N	β
A7	41.011	0.7513
A5	47.708	0.7083
A3	42.933	1.0741
A2	46.159	1.2098

design combination matched the optimum design. However, the graph also shows that at 1100 °C, the original design combination is far from optimum and there is great variation of operation. Hence, the system was not robust.

Table 18.9 Result of confirmation run: (a) A7; (b) A5; (c) A3; (d) A2

A7	Prediction	Confirmation	% Difference
Optimum S/N Design	41.011	52.013	26.8%
Initial S/N Design	25.900	25.9	0.0%
Gain	15.111	26.113	72.8%
Sigma reduction	82.5%	95.1%	

(a)

A5	Prediction	Confirmation	% Difference
Optimum S/N Design	47.708	51.036	7.0%
Initial S/N Design	25.900	25.9	0.0%
Gain	21.808	25.136	15.3%
Sigma reduction	91.9%	94.5%	

(b)

A3	Prediction	Confirmation	% Difference
Optimum S/N Design	42.933	42.458	1.1%
Initial S/N Design	25.900	25.9	0.0%
Gain	17.033	16.558	2.8%
Sigma reduction	86.0%	85.2%	

(c)

A2	Prediction	Confirmation	% Difference
Optimum S/N Design	46.159	42.020	9.0%
Initial S/N Design	25.900	25.9	0.0%
Gain	20.259	16.120	20.4%
Sigma reduction	90.4%	84.5%	

(d)

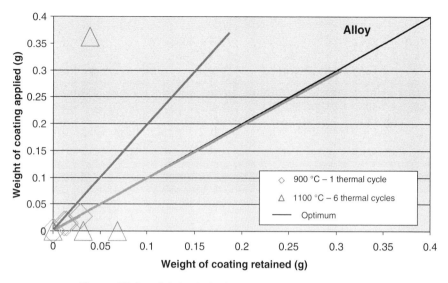

Figure 18.8 Original design used for the reformer.

Figure 18.9 is the design improvement of the washcoat adhesion affected by robust engineering methods. The two step process both narrowed the variation in the response, made the system more robust to both variations in temperature and thermal cycling and brought the response of the washcoat adhesion close to ideal behavior.

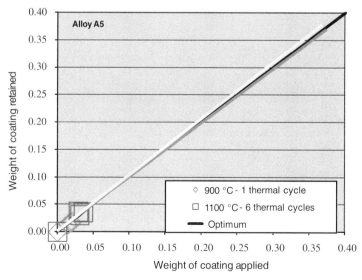

Figure 18.9 The design parameter combination for the optimal design.

Table 18.10 Measurement system evaluation

Sample	$y_{i(m1)}$	$y_{i(m2)}$	y-bar$_{sample}$	$e_{i(mj)} = y_{i(mj)} -$ y-bar$_{sample}$	$e_i^{2(mj)}$	
1	1.547	1.546	1.5465	0.0005	−0.0005	0.0000002	0.0000002
2	1.870	1.870	1.8700	0	0	0.0000000	0.0000000
3	1.840	1.840	1.8400	0	0	0.0000000	0.0000000
4	1.812	1.812	1.8120	0	0	0.0000000	0.0000000
5	1.829	1.829	1.8290	0	0	0.0000000	0.0000000
6	1.772	1.772	1.7720	0	0	0.0000000	0.0000000
7	1.794	1.793	1.7935	0.0005	−0.0005	0.0000003	0.0000002
8	1.764	1.764	1.7640	0	0	0.0000000	0.0000000
9	1.878	1.877	1.8775	0.0005	−0.0005	0.0000002	0.0000002
10	1.754	1.753	1.7535	0.0005	−0.0005	0.0000003	0.0000002

y-bar$_{mj}$ = 1.786   Sum$[(e_{i(mj)}^2)]$ = 0.0000020

s_{mj} = 0.0918 s_m = 0.000447214

18.9 Measurement System Evaluation

Table 18.10 gives the measurement system evaluation and Figure 18.10 gives the minimum detectable difference.

18.10 Conclusion

The principal lesson ascertained from this study is the importance of the choice of substrate composition to washcoat adhesion. Substrate composition is the dominate control factor for washcoat adhesion. If the substrate composition is chosen poorly, none of the other control factors could mitigate the

Minimum Detectable Difference

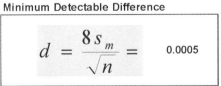

$$d = \frac{8\,s_m}{\sqrt{n}} = \quad 0.0005$$

n is the number of treatment combinations in the experiment
In our case we had 54 combinations, so n = 54

Figure 18.10 Minimum detectable difference.

detrimental effects. The surface preparation came a distance second as an influential control factor. All other control factors had far less influence than substrate composition and surface preparation.

The study was confirmed using four predictions. Alloy compositions A5, A2 and A3 were confirmed. These compositions confirmed within 10% of prediction. However, alloy composition, A7 did better than predicted. A7 confirmed at 26%. This was possibly due to an interaction within the study.

This study produced an increase in washcoat adhesion from less than 15 hours of operation to more than a 100 hours of operation. Subsequent, robust engineering studies have enhanced washcoat adhesion on metallic substrates even further to greater than 800 hours. It should be noted that the most recent study was stopped at 800 hours due to equipment timing issues rather than failure of the washcoat adhesion. This progress in washcoat adhesion from 2003 to present is shown in Figure 18.11. This growth in knowledge and experience in washcoat adhesion on metallic substrates was achieved by the application of Taguchi Robust Engineering Methods. The information from this study has been used in currently operating reformers,

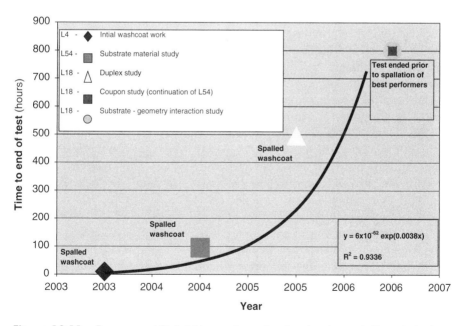

Figure 18.11 Progress at Delphi in washcoat adhesion to metallic substrates through application of robust engineering methods to date.

which have shown thousands of hours of operation without excessive spallation of the washcoat.

It must be noted that the cleaning methods used in this study were among the best possible. Some very bad choices for cleaning methods could be employed which would give dramatically different results than those observed in this study. The flat behavior of the signal-to-noise ratio shown in this study should not lead the reader to believe that cleaning methods are unimportant. But, merely that this study was trying to determine the best method from already excellent choices.

This study was useful in capturing and documenting the performance of a wide range of metallic substrate materials, and has served as a reference in the selection of metallic materials for reformer designs.

18.11   Supplemental Background Information

The reformer of a solid oxide fuel cell, SOFC system converts liquid and gaseous fuels into a hydrocarbon gas, termed "reformate." Reformate, for an SOFC system, is ideally a gaseous mixture containing: hydrogen, H_2; carbon monoxide, CO; carbon dioxide, CO_2; nitrogen, N_2; and water vapor, H_2O. Nonideally, the reformate gas could contain higher hydrocarbons such as ethylene-C_2H_4, etc. These higher hydrocarbons pose a danger to an SOFC system due to their tendency to decompose on hot surfaces including the reformate delivery path to the SOFC stack or within the SOFC stack itself. Such decomposition is termed pyrolysis, coking or hydrocarbon decomposition. The formation of carbon before the stack or within the stack will eventually lead to a condition called "concentration polarization" which reduces the power output of a fuel cell and, consequently, the SOFC stack. Concentration polarization occurs when the feed gases for an electrochemical reaction are blocked from reaching the catalytic reaction site. This blocking of the catalytic reaction sites causes a decrease in the rate of the electrochemical reaction and, therefore, negatively affects the power output of an SOFC system. The production of unacceptable reformate, that is, reformate with a high coking potential, is directly related to the amount of catalyst used in the reformer. The more catalyst available in the reformer; the higher the quality of reformate that is produced. If there is not enough catalyst available, then significant amounts of higher hydrocarbons can be produced which will then form carbon by pyrolysis. Using more and more catalyst to accommodate eventual loss of the catalyst during operation is not economically feasible.

Table 18.11 is a blank L54 orthogonal matrix with 18 levels in column one. Columns 2 and 9–14 are used to make the 18-level column.

Table 18.11 Blank L54 orthogonal matrix with 18 levels in column 1 (columns 2 and 9–14 are used to make the 18-level column)

No.	1*	3	4	5	6	7	8	15	16	17	18	19	20	21	22	23	24	25
1	1	1	1	1	1	1	1	1	1	1	1	1	1	1	1	1	1	1
2	2	1	1	1	1	1	1	2	2	2	2	2	2	2	2	2	2	2
3	3	1	1	1	1	1	1	3	3	3	3	3	3	3	3	3	3	3
4	1	2	2	2	2	2	2	2	3	2	3	2	3	2	3	2	3	2
5	2	2	2	2	2	2	2	3	1	3	1	3	1	3	1	3	1	3
6	3	2	2	2	2	2	2	1	2	1	2	1	2	1	2	1	2	1
7	1	3	3	3	3	3	3	3	2	3	2	3	2	3	2	3	2	3
8	2	3	3	3	3	3	3	1	3	1	3	1	3	1	3	1	3	1
9	3	3	3	3	3	3	3	2	1	2	1	2	1	2	1	2	1	2
10	4	1	1	2	2	3	3	1	1	1	1	2	3	2	3	3	2	3
11	5	1	1	2	2	3	3	2	2	2	2	3	1	3	1	1	3	1
12	6	1	1	2	2	3	3	3	3	3	3	1	2	1	2	2	1	2
13	4	2	2	3	3	1	1	2	3	2	3	3	2	3	2	1	1	1
14	5	2	2	3	3	1	1	3	1	3	1	1	3	1	3	2	2	2
15	6	2	2	3	3	1	1	1	2	1	2	2	1	2	1	3	3	3
16	4	3	3	1	1	2	2	3	2	3	2	1	1	1	1	2	3	2
17	5	3	3	1	1	2	2	1	3	1	3	2	2	2	2	3	1	3
18	6	3	3	1	1	2	2	2	1	2	1	3	3	3	3	1	2	1
19	7	1	2	1	3	2	3	1	1	2	3	1	1	3	2	2	3	3
20	8	1	2	1	3	2	3	2	2	3	1	2	2	1	3	3	1	1

Table 18.11 (*Continued*)

No.	1*	3	4	5	6	7	8	15	16	17	18	19	20	21	22	23	24	25
21	9	1	2	1	3	2	3	3	3	1	2	3	3	2	1	1	2	2
22	7	2	3	2	1	3	1	2	3	3	2	2	3	1	1	3	2	1
23	8	2	3	2	1	3	1	3	1	1	3	3	1	2	2	1	3	2
24	9	2	3	2	1	3	1	1	2	2	1	1	2	3	3	2	1	3
25	7	3	1	3	2	1	2	3	2	1	1	3	2	2	3	1	1	2
26	8	3	1	3	2	1	2	1	3	2	2	1	3	3	1	2	2	3
27	9	3	1	3	2	1	2	2	1	3	3	2	1	1	2	3	3	1
28	10	1	3	3	2	2	1	1	1	3	2	3	2	2	3	2	3	1
29	11	1	3	3	2	2	1	2	2	1	3	1	3	3	1	3	1	2
30	12	1	3	3	2	2	1	3	3	2	1	2	1	1	2	1	2	3
31	10	2	1	1	3	3	2	2	3	1	1	1	1	1	2	3	2	2
32	11	2	1	1	3	3	2	3	1	2	2	2	2	2	3	1	3	3
33	12	2	1	1	3	3	2	1	2	3	3	3	3	3	1	2	1	1
34	10	3	2	2	1	1	3	3	2	2	3	2	3	1	1	1	1	3
35	11	3	2	2	1	1	3	1	3	3	1	3	1	2	2	2	2	1
36	12	3	2	2	1	1	3	2	1	1	2	1	2	3	3	3	3	2
37	13	1	2	3	1	3	2	1	1	2	3	3	2	1	1	3	2	3
38	14	1	2	3	1	3	2	2	2	3	1	1	3	2	2	1	3	1
39	15	1	2	3	1	3	2	3	3	1	2	2	1	3	3	2	1	3
40	13	2	3	1	2	1	3	2	3	3	2	1	1	2	3	1	1	3

Run																	
41	2	3	1	2	1	3	1	1	3	3	1	2	2	2	2	2	14
42	2	3	1	1	1	3	1	2	1	3	1	2	2	3	3	3	15
43	3	1	2	1	2	3	1	1	3	3	3	2	2	2	2	3	13
44	3	1	2	1	2	3	1	1	3	3	3	2	2	2	3	3	14
45	3	1	2	1	2	3	1	3	1	3	3	2	3	3	1	1	15
46	1	2	2	3	1	2	2	3	1	1	3	3	3	1	1	1	16
47	1	2	2	3	2	2	2	3	1	1	3	3	3	1	1	1	17
48	2	3	2	3	3	2	2	3	1	1	3	3	3	1	3	1	18
49	2	3	1	1	2	3	1	1	2	3	1	1	1	3	1	2	16
50	2	1	2	2	3	1	2	2	2	3	2	1	1	3	1	2	17
51	2	2	3	3	3	3	3	3	2	3	2	2	1	3	1	2	18
52	1	3	1	3	2	1	1	2	3	1	3	3	1	2	2	3	16
53	2	1	3	1	3	2	2	3	3	1	3	3	1	2	2	1	17
54	3	2	2	1	1	3	3	1	1	2	1	3	1	2	2	2	18

18.12   Acknowledgment

The author would like to thank Chris Mergler, Anthony DeRose, Greg Callahan and Anne Mikels for all the work to run and complete the analysis of this large study. This work could not have been completed without their assistance. Thanks to Alan Wu and Craig D. Smith of ASI Consulting Group, LLC, USA for their coaching of this study. Special thanks to Steven Shaffer for his support of this research. Few bosses possess the vision to see the inestimable value of such a comprehensive body of work.

18.13   Reference and Further Reading

1. Gaskell, D. (1981) *Introduction to Metallurgical Thermodynamics*, New York, Washington, Philadelphia, London: Hemisphere Publishing Corporation.
2. *Robust Engineering Workshop Manual*, ASI Consulting Group, LLC, www.asiusa.com, 2001.
3. England, D.M., and Virkar, A.V. (2001) *Journal of Electrochemical Society*, **148** (4): A330.

This case study is contributed by Diane England of Delphi Automotive Systems, USA.

19

Making Better Decisions Faster: Sequential Application of Robust Engineering to Math-Models, CAE Simulations, and Accelerated Testing

Robert Bosch Corporation, USA

19.1 Executive Summary

Enclosures for automotive power electronics have thermal loads (TLs) and boundary conditions (BCs) which are fluid and ambiguous. A three phase study was initiated to integrate these TLs and BCs noise factors into our thermal math-models, simulations, and accelerated tests to assess the thermal robustness of our enclosures.

Robust Optimization: World's Best Practices for Developing Winning Vehicles,
First Edition. Subir Chowdhury and Shin Taguchi.
© 2016 Subir Chowdhury, Shin Taguchi, and ASI Consulting Group, LLC.
Published 2016 by John Wiley & Sons, Ltd.

This project (the first phase of the study) uses a math model to target the following goals:

- Create thermal robustness design template for making better decisions faster.
- Determine critical design parameters and their levels for thermal robustness.
- Cascade thermal robustness learning from math-model into CAE and accelerated testing phases.
- Test Taguchi Robust Engineering methodology on thermal math model.

The thermal equivalent circuit's closed-form solution generated 1,296 data points inside the six signal level, L-18 inner by L-12 noise matrix. The Taguchi Robust Engineering methodology clearly identified the set of critical design parameters which will be implemented in CAE and accelerated testing.

19.2 Introduction

As the cost of solid-state switches continues to decrease and their respective performance increases, the application of electronic devices to control automotive electrical devices is growing at an exponential rate. Most devices are soldered on to printed circuit boards (PCBs) and enclosed in either a plastic or a metal enclosure. The enclosure material decision is frequently determined by the heat transfer requirements of the electronics.

All electronic devices, especially solid-state switching devices such as the Field-Effects-Transistors (FETs), generate heat during their operation. The thermal energy has to either be absorbed by the device, the PCB, and the enclosure or be transferred to the surrounding environment. The effectiveness of this heat transfer will dictate what temperature the electronics will operate at during steady state operation.

There are two primary challenges in automotive applications. First, the thermal loads frequently have a large degree of uncertainty to them. These loads are sensitive to duty cycle assumptions, battery voltage variation, aging, and the temperature of the device. The temperature of the device is also affected by the total device heat flux and overall heat transfer performance.

The second challenge is the thermal boundary conditions that are poorly defined or unknown until after the product's design is required to be complete. The packaging orientation of the control unit, the ability of the air around the unit to dissipate into the vehicle and the temperature of surrounding components all have a significant impact on the heat transfer from the electronics module. The automotive Original Equipment Manufacturers (OEMs) typically specify the operating temperature, the packaging envelope, and the air movement assumptions. For example, 85 °C under still air conditions in a package envelope of 90 mm × 120 mm × 35 mm.

The conventional approach to these challenges is to design the enclosure to deliver sufficient heat transfer under the "worse-case" thermal loads and boundary conditions, which will maintain the electronics below an absolute temperature threshold. Historically, a multitude of technical knowledge and tools (software and hardware) has been applied to enclosure's design so the design can perform under these extreme thermal loads and boundary conditions.

Unfortunately, over-designed products are no longer competitive in the automotive sector. In addition, there has been limited progress by the OEMs to eliminate, reduce, or control these thermal noise factors. Designing thermally robust enclosures appears to be the logical strategy for future OEM products.

19.2.1 Thermal Equivalent Circuit – Detailed

A thermal equivalent circuit is a mathematical approximation of the heat transfer in an object. The mathematical approximation is in the form of an electrical circuit, which can be solved using Kirkoff's and Ohm's laws. Resistors can approximate thermal impedances for conduction, convection, and radiation. Conductors (inverted resistors) approximate thermal capacitance. Heat flow is represented by current and the change in temperature is the voltage potential. The full detailed thermal equivalent circuit for the reference body control unit in shown Figure 19.1.

19.2.2 Thermal Equivalent Circuit – Simplified

For this project (the first phase of our study), the circuit was simplified using standard formulations for series and parallel resistors. The analysis was further simplified by eliminating the transient component and only evaluating the circuit under steady state conditions. In terms of heat transfer, the thermal capacitances of the electronic devices, the PCB, and the enclosures were ignored. Figure 19.2 depicts this simplified circuit.

19.2.3 Closed Form Solution

The circuit outlined in Figure 19.2 yields the following equations. These simulations equations were solved using basic algebraic matrix techniques.

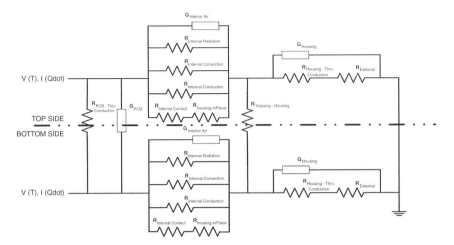

Figure 19.1   Detailed transient thermal equivalent circuit

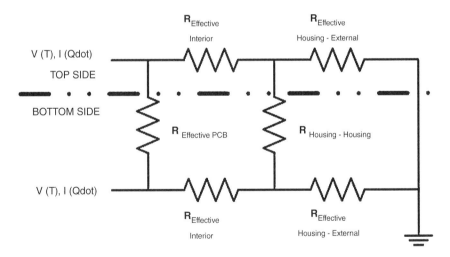

Figure 19.2   Simple steady state thermal equivalent circuit

The resulting closed form equations were integrated into an electronic spreadsheet for evaluation.

$$R_2 \cdot i_2 + R_4 \cdot i_4 - (R_1 \cdot i_1 + R_3 \cdot i_3) = 0 \quad (i_3 + i_4) - i_6 = 0$$
$$R_5 \cdot i_5 - (R_4 \cdot i_4 + R_6 \cdot i_6) = 0 \quad i_2 - (i_4 + i_5) = 0$$
$$i_1 + i_2 = i_{cs} \quad i_3 - i_1 = i_{ss} \quad i_5 + i_6 = i_{cs} + i_{ss}$$

19.3 Objective

One challenge with any new development approach is applying it with the maximum benefit and the minimum disruption. For this reason, this ASI Taguchi DFSS IDDOV type subsystem development was intentionally decoupled from the production design and applied to a new module concept. However, this analysis uses production noise (thermal loads, and boundary conditions) and similar physical characteristics so that development engineers on the production project can learn from any insights (make better decisions) without being hindered by the unknown progress of the Taguchi approach.

The enclosure's robustness to thermal noise factors was chosen for two reasons. First, high temperatures and thermal cycling have a severe negative impact on electronics reliability. Second, automotive power electronics typically have fluid and ambiguous thermal loads (TLs) and boundary conditions (BCs). As a result, thermal robustness is one of the critical concerns for the emulated and future production designs.

A three-phase study was initiated to challenge our thermal management design decisions for electronic control modules using the Taguchi Robust Engineering methodology. The goals of this project are to use math-model based on our thermal equivalent circuit as follows:

19.3.1 Thermal Robustness Design Template

In our lean product development efforts, we realized that certain decisions are always present in a project although the application details can vary dramatically. One set of decisions include those relating to the mechanical configuration of a control module. Before specific decisions can be made about geometry, the more general decisions need to be made about the enclosure (e.g., should the enclosure be metal?). The advantage of the math-model has always been its ability to explore the various configurations and options quickly. One project goal is to develop a thermal robustness template to help design engineers make better high-level thermal management decisions in less time and with fewer resources. By simply changing the factor levels and following the flow of the template, the consistency to the thermal management decisions should increase (i.e., with less engineer-engineer variance).

19.3.2 Critical Design Parameters for Thermal Robustness

One tendency of engineers is to design enclosures for thermal performance only. As a result, each time the noise factors change or become more refined the design is compromised unless it is retuned. Retuning or changing a design, which is optimized for thermal performance, only becomes more difficult as the project approach production and frequently forces a completely different configuration or an expensive band-aid.

By focusing on robustness and performance in that sequence (Taguchi two-step optimization), the expectation is that the design parameters (control factors) and their levels will only be refined as the thermal noise factors stabilize and become more accurate. Major design changes should be less likely because of the robust thermal design decisions. Decisions should be better because they are based on the ideal function of the device and tested against noise factors set to their extreme levels. The control factors and their respective levels will be selected to be robust to these conditions.

19.3.3 Cascade Learning (aka Leveraged Knowledge)

The next goal is to integrate the thermal management decisions made about the design parameters into 3D computer models. The 3D models and the leveraged knowledge from the math-model should enable more precise computer-assisted engineering (CAE) simulations and physical testing to be more effectively employed. Being able to reduce and focus the scope of this subsequent engineering work provides a significant business opportunity since CAE and testing are resource intensive endeavors. If the number of factors and or levels can be reduced, or complex noise arrays are replaced with a more efficient compounded noise strategy (e.g., N1, N2) then more can be learned about the design with the same resources. We should be able to make better decisions faster from what we simulate and/or test.

19.3.4 Test Taguchi Robust Engineering Methodology

Math-models should provide an opportunity to test your understanding of Taguchi Robust Engineering methodology and its applicability to your project. Their results should make sense even if they were not expected. Math-models should also enable different type of responses, factors, and levels to be explored in order to yield the highest degree of insight. In addition, because of the simplicity of most math-models, the verification of the predicted optimal or tuned design should be relatively simple with CAE or accelerated testing.

19.4 Robust Optimization

The ASI robust engineering eight-step process for optimizing designs using computer simulations was followed. The closed form equations from the thermal equivalent circuit were integrated into an electronic spreadsheet to create the simulation.

19.4.1 Scope and P-Diagram

The scope of the ideal function is the heat transfer between the electronics inside the control unit and the surrounding vehicle environment as shown in Figure 19.3. From this scope, a detailed P-diagram was created and partially shown in Figure 19.4. This P-diagram was the basis for the math-model, and the noise, signal and control factors strategies.

19.4.2 Ideal Function

The thermal transfer ideal function for the enclosure of an electronic device is thermal energy transportation from the operating electronics to its passive surrounding environment. The flow of thermal energy (heat) is typically indirectly measured by a device's change in temperature (dT).

Specific to electronics, the dT on the surface of the PCB is inversely related to the amount of thermal energy transferred to its surrounding environment from the operating electronics. This dT or temperature potential is analogous to voltage potential in electrical circuits when they are used to approximate the thermal performance of a device. Thermal energy or power (heat flux) is analogous to the electrical current. The ability of the matter in and around the device to contribute to the thermal energy transportation in terms of

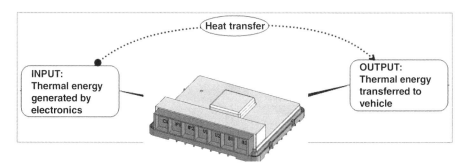

Figure 19.3 Scope

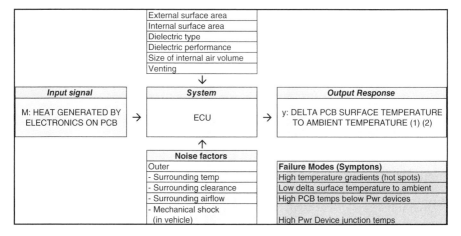

Figure 19.4 P-diagram (cropped for brevity)

conduction, convection, and radiation is analogous to resistance or impedance. The ideal function based on dT is shown in Figure 19.5.

The Smaller-the-Better (STB) nondynamic signal-to-noise (S/N) ratio approach is the typical choice for a symptomatic response like dT. The goal, however, was to challenge our thinking and decisions on all thermal management technologies. Understanding robustness dynamically would therefore be more valuable.

Changing the dT response to the signal and the heat flux signal to the response should give a better indication of the efficiency of the heat transfer. Figure 19.6 shows how the ideal function would respond to increasing Beta.

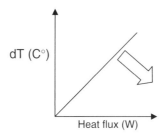

Figure 19.5 Ideal function – dT response

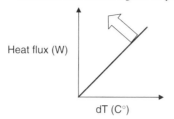

Maximize heat flux for a given dT potential

Heat flux (W)

dT (C°)

Figure 19.6   Ideal function heat flux response

19.4.3 Signal and Noise Strategy

The simulation was run in an electronic spreadsheet using a standard orthogonal L-12 array for the noise factors replicated for each of the six signal levels. The noise factors were derived from the thermal equivalent model, and the respective equations defining each thermal resistance. Variables were determined to be noise factors if they were fluid (changing), ambiguous (insufficiently specified) or not the enclosure design engineer's decisions. The levels for these factors were set to their plausible extremes for the specific product being evaluated. Table 19.1 outlines the noise factors and the levels. Within the math-model these levels were affected by the relevant variables in the closed form solution.

Table 19.1   Noise factors

ID	NOISE FACTORS	Level 1	Level 2
N	Convection film coefficient of exterior surface	Low	High
P	Areas around ECU that air is allowed to flow	Btm	All
R	External clearance around ECU	2	20
S	Size of PCB	145X145	155X155
T	PCB construction	2 L	4 L
U	Average % copper per layer	60	95
V	% Thermal power on top side of PCB	55%	65%
W	Special 1	No	Yes
X	Orientation of ECU	Flat	Side
Y	Special 2	Corner	Edge
Z	PCB – Hsg Contact thermal impedance cm2- K/W	Low	High

19.4.4 Input Signal

The total thermal power generated by the electronics (heat flux) was chosen as the input signal for this project. The math-model proportioned this total thermal power proportioned to each side of the PCB. The total thermal power ranged from 10 W to 15 watts in 1 W increment to cover the expected range.

The affect of having the dT potential as the signal was accomplished with the same data generated from the math-model using the heat flux as the signal. This is a reasonable approach since a unique heat flux value was generated for each of the 72 responses per inner array configuration.

19.4.5 Control Factors and Levels

A standard orthogonal L-18 matrix was used for eight control factors in the inner array. Though the selection of the control factors started with the P-diagram, the final set of factors was determined by the math-model variables, which represented typical module choices. All modes of thermal transfer were covered by these factors (conduction, convection, and radiation).

The levels of the control factors represent the typical and plausible limits for the product under study. In the math-model, most of the factor levels were converted to numerical equivalents (e.g., the material type became a specific thermal conductivity values). However, some factors were represented by multiplication factors (e.g., heat sink features levels were factors multipliers of the external surface area). Table 19.2 shows the control factors and their levels.

Table 19.2 Control factors

ID	Control factor	Level 1	Level 2	Level 3
A	Emissivity	Level 1	Level 2	-
B	Material type	Type 1	Type 2	Type 3
C	Material thickness	Level 1	Level 2	Level 3
D	Heat sink features	Type 1	Type 2	Type 3
E	Internal convection – top	Level 1	Level 2	Level 3
F	Internal convection – bottom	Level 1	Level 2	Level 3
G	Exchanging	Type 1	Type 1	Type 1
H	Contact area	Type 1	Type 2	Type 3

19.4.6 Math-Model Generated Data

The electronic spreadsheet calculated 23 variables for each node on the 18 x 12 x 6 matrix. This created 29 808 calculated cells in the electronic spreadsheet and generated 1,296 dTs for each side of the printed circuit board.

For simplicity, the details of these calculations and their resulting numerical values are omitted except as necessary to demonstrate certain calculations. However, Figure 19.7 plots the raw top side PCB dT data for all of the 18 runs and all the associated noise and signal levels.

19.4.7 Data Analysis

For each side of the PCB the data was analyzed first using the nondynamic delta temperature responses and then with the dynamic deviation from the maximum delta temperature response. Thermal robustness was assessed using S/N and subsystem thermal resistance was assessed by the associated Beta or slope of the data. To demonstrate the calculations of S/N and Beta for the dynamic and nondynamic responses Table 19.3 and Table 19.4 show some of the raw data.

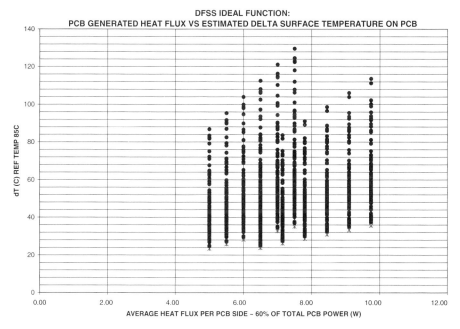

Figure 19.7 Plot of ideal function raw data

Table 19.3 dT response

	Signal level	NOISE ARRAY CONFIGURATION											
		1	2	3	4	5	6	7	8	9	10	11	12
1	M1	79.14	68.52	64.55	61.44	75.08	57.68	53.10	61.74	57.61	62.23	58.14	57.11
2	M1	60.73	53.43	49.84	48.13	57.68	45.15	42.85	49.48	45.66	48.87	46.19	44.42
3	M1	53.42	47.73	44.14	43.17	50.75	40.23	39.40	45.17	41.21	43.56	42.00	39.07
4	M1	81.76	76.07	72.22	64.00	83.38	59.84	59.12	62.46	59.22	63.31	64.84	62.79
5	M1	46.18	49.24	46.36	42.56	50.38	34.97	37.65	42.39	34.33	41.94	42.60	38.91
6	M1	32.11	32.35	30.80	26.35	34.67	23.78	25.27	25.69	23.40	26.03	27.81	26.92
7	M1	49.19	47.54	45.09	38.47	51.87	35.98	37.24	37.59	35.66	38.33	40.64	39.75
8	M1	39.62	41.64	39.50	36.72	41.93	29.46	30.66	36.54	28.98	36.52	36.03	32.33
9	M1	45.77	44.94	43.10	38.08	47.53	34.14	34.47	36.53	33.15	36.71	38.48	36.62
10	M1	39.69	35.49	32.94	32.06	37.84	30.22	28.44	32.75	30.07	32.35	30.38	29.99
11	M1	86.74	74.43	71.08	66.97	82.28	62.77	57.14	66.29	62.66	67.62	63.32	61.94
12	M1	48.43	42.68	40.29	38.47	46.31	36.36	33.90	38.92	36.41	39.27	36.79	36.35
13	M1	48.23	45.76	43.07	39.07	49.24	36.34	35.88	38.22	35.56	38.18	38.85	38.11
14	M1	34.25	32.72	30.93	27.03	35.57	25.32	25.57	26.46	24.97	26.85	27.84	27.58
15	M1	57.61	54.30	51.96	48.66	57.58	42.22	40.63	47.65	41.64	48.29	46.36	43.66
16	M1	33.63	33.13	31.28	27.25	35.72	25.09	26.07	26.78	24.73	27.06	28.33	27.87
17	M1	52.99	50.29	47.87	42.43	54.39	39.52	39.11	40.72	38.38	40.89	42.61	41.87
18	M1	48.80	45.93	43.82	41.26	48.14	35.47	33.60	40.55	35.02	41.13	39.04	36.31
18	M2	53.67	50.48	48.18	45.36	52.93	39.01	36.94	44.57	38.51	45.21	42.91	39.93
18	M3	58.55	55.03	52.53	49.45	57.73	42.56	40.29	48.59	42.01	49.29	46.78	43.54
18	M4	63.42	59.58	56.87	53.54	62.52	46.10	43.63	52.60	45.50	53.37	50.65	47.16
18	M5	68.29	64.12	61.21	57.63	67.31	49.64	46.96	56.60	48.99	57.43	54.51	50.77
18	M6	73.16	68.66	65.55	61.71	72.10	53.17	50.30	60.61	52.49	61.50	58.36	54.37

INNER ARRAY CONFIGURATION

Table 19.4 Heat flux (W) signal

SIGNAL INPUT TOTAL HEAT FLUX (W)		PCB HEAT FLUX (W) PER NOISE CONFIGURATION											
		1	2	3	4	5	6	7	8	9	10	11	12
M1	10	5.00	6.50	5.00	6.50	5.00	6.50	6.50	6.50	5.00	5.00	5.00	6.50
M2	11	5.50	7.15	5.50	7.15	5.50	7.15	7.15	7.15	5.50	5.50	5.50	7.15
M3	12	6.00	7.80	6.00	7.80	6.00	7.80	7.80	7.80	6.00	6.00	6.00	7.80
M4	13	6.50	8.45	6.50	8.45	6.50	8.45	8.45	8.45	6.50	6.50	6.50	8.45
M5	14	7.00	9.10	7.00	9.10	7.00	9.10	9.10	9.10	7.00	7.00	7.00	9.10
M6	15	7.50	9.75	7.50	9.75	7.50	9.75	9.75	9.75	7.50	7.50	7.50	9.75

19.4.8 Thermal Robustness (Signal-to-Noise)

The first step in the Taguchi two-step optimization is to focus on the variation. Temperature is the typical measurement for thermal performance, but it is a symptomatic nonenergy based response. An energy-based response is preferred to evaluate robustness. The change in temperature is proportional to the energy stored by the system, which is directly related to how much thermal energy is transferred from the electronics to the surrounding environment. The nondynamic Smaller-the-Better (STB) S/N approach was chosen since we want to minimize the amount of stored energy. Equation (19.1) shows how the STB S/N was calculated for the dTs.

$$S/N = 10 \log \frac{1}{\frac{1}{n} \sum_{i=1}^{n} y_i^2} \tag{19.1}$$

where y is the delta temperature and n = 72 (12 X 6)

Using data from Table 19.3, the S/N for the 18th inner configuration equals −34.3.

From Figure 19.8, the STB S/N analysis indicates that factor D (Heat Sink features) has the largest impact on robustness across all noise and signal levels with a range of 4.4 db. The S/N range for factors B (Material type) and G (Exchanging) was 2.0 db and 1.4 db respectively. Interestingly though most of the S/N gain for material type was achieved by moving from level 1 to 2 (1.4 db) and for the "Exchanging" from levels 3 to 2 (1.6 db). This data suggests that either level 2 or 3 for "Material type" and levels 1 or 2 for "Exchanging" are acceptable robustness settings. Factors A (Emissivity) and C (Material thickness) did show some differential between the levels, but less than 1 db. A potential cost savings may exist.

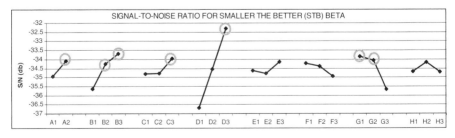

Figure 19.8 Response plot – nondynamic STB S/N (dT)

Next, the dynamic S/N of for these delta temperatures were calculated using the "Most Common Case" formulation as outlined below in Equation (19.2).

$$S/N = 10 \log\left(\left(\frac{1}{r}\right) \left(\frac{(S_\beta - V_e)}{V_e}\right) \right)$$ (19.2)

Variation was estimated by Equation (19.3):

$$V_e = \frac{S_T - S_\beta}{n - 1}$$ (19.3)

where

$$S_T = \sum_{i=1}^{n} y_i^2$$ (19.4)

$$S_\beta = \frac{\left(\sum_{i=1}^{n} M_i y_i\right)^2}{r}$$ (19.5)

$$r = \sum_{i=1}^{n} M_i^2$$ (19.6)

Using the dTs (y), and the specific thermal power (M) from Table 19.3, the S/N for 18th configuration of the inner array was −3.84.

The overall magnitude of the S/N's was reduced going from nondynamic to dynamic analysis as shown in Figure 19.9. In addition, the total range of S/N was smaller (< 0.6 db). More troublesome though was the incongruence between this response plot (Figure 19.9) and the STB S/N response plot (Figure 19.8) which was more inline with our engineering experience.

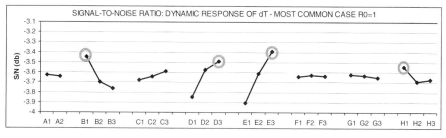

Figure 19.9   Response plot – dynamic S/N (dT) – most common case

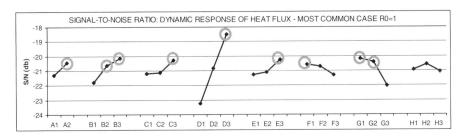

Figure 19.10 Response plot – dynamic S/N (Q) – most common case

Inverting the signal (heat flux) and the response (dT) provides an alternative method of looking at the data. For a temperature potential above the ambient temperature (dT) how much heat can be flowed through the system. Figure 19.10 shows the dynamic S/N (Most Common Case) when dT is the signal and the heat flux is the response. This response plot is more consistent with the STB S/N response plot (Figure 19.8).

19.4.9 Subsystem Thermal Resistance (Beta)

The Beta of the ideal function is equivalent to the subsystem thermal resistance. Again several methods to analysis it were evaluated.

First, the "Most Common Case" formulation of the dynamic response the Beta was calculated as shown in Equation (19.7).

$$\beta = \sum_{i=1}^{n} \frac{M_i y_i}{r} \tag{19.7}$$

where

$$r = \sum_{i=1}^{n} M_i^2 \tag{19.8}$$

The response plot for thermal resistance, Figure 19.11, was congruent (albeit inverted) to the STB S/N response plot (Figure 19.8).

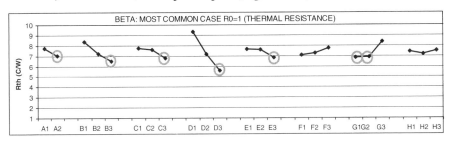

Figure 19.11 Response plot – dynamic beta – most common

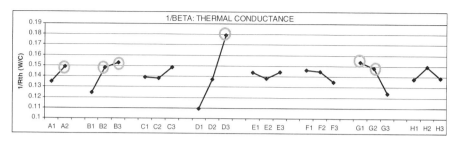

Figure 19.12   Response plot – dynamic 1/beta – most common case

This thermal resistance response plot (Figure 19.11) however does not provide any insight into the variability of the thermal resistance of any factor at any level. If robustness was not evaluated before performance then most likely factors A (Emissivity) and C (Material thickness) would not be important design factors even though they have a substantial impact on thermal robustness (Figure 19.9 and Figure 19.12). In addition, the less expensive level 2 material types (Factor C) would not be considered because their poor performance relative to level 1 material types. This is important because using Type 2 materials have the potential to reduce costs when used as an alternative to Type 3 materials. The Taguchi two-step optimization enabled the realization that Type 2 is a viable option because of its similar thermal robustness to Type 3 and the thermal resistance could be tuned by another more effective factor such as the heat sinking (Factor D).

Figure 19.11 can be confusing when contrasting its optimal setting with the optimal setting for thermal robustness in the S/N response plots (Figure 19.8 and Figure 19.10). The directions of optimal levels are inverted from the S/N analysis. The simple solution is to invert the calculated beta or convert the thermal resistance into the thermal conductance. Now the response plot as shown in Figure 19.12 matches the pattern of the S/N response plots and is more congruent with those results.

19.4.10   Prediction and Confirmation

The 11th configuration of the inner array closely resembled the configuration of the intended production design or the "Reference" design. The "Optimal" design has control factors set to their highest S/N levels. The "Tuned" design has the control factors set the best compromise between thermal robustness, thermal resistance, and implementation cost. Table 19.5 below summarizes the setting.

First, the raw data was analyzed. Figure 19.13 shows the dT response against the thermal power for all 216 noise and signal combination for the

Table 19.5 Setting for reference, optimal and tuned designs

		CONTROL FACTORS (INNER ARRAY)							
		A	B	C	D	E	F	G	H
		Emmissivity	Material	Thickness	Heat sink features	Internal clearance (T)	Internal clearance (B)	Venting	Contact area
Design	Reference	2	1	2	1	1	3	3	2
	Optimal	2	3	3	3	1	1	1	2
	Tuned	2	2	2	3	2	2	2	3

Reference design only. The ellipse drawn around all the data denotes the variation in response and the best-fit line through zero approximates the Beta.

Figure 19.14 shows the impact to the ideal function when the parameters are set to their optimal setting for S/N. Since the axis of both plots (Figure 19.13 and Figure 19.14) have the same limits, it becomes readily apparent how the thermal robustness is improved (ie variation reduced) and thermal resistance is lower (i.e., Beta lowered).

The impact of tuning on the ideal function plot is harder to discern (nearly identical) when the "Optimal" design and "Tuned" design (Figure 19.15) are compared.

As a further check, the average signal levels were plotted against the average dT for each set of 12 noise conditions. Figure 19.16 shows the relative average thermal resistance (Beta) of the "Reference," the "Optimal" and the "Tuned" designs.

Table 19.6 summarizes the predicted S/Ns and Betas and their confirmations. In this table, the subtle difference between the "Optimal" and "Tuned" designs can be detected. Although the thermal robustness of "Tuned" design is 2.0–2.4 db lower than the "Optimal" design, it is more than 7 db higher the "Reference" design.

As expected the Beta for dT show that the "Tuned" performs less effectively than the "Optimal" (~6% higher thermal resistance). However, the improvement in performance from the "Reference" to the "Tuned" designs is more significant (~63% reduction in thermal resistance).

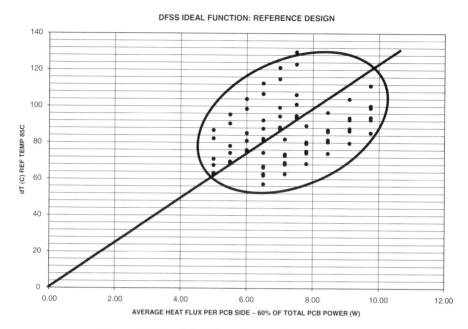

Figure 19.13   Ideal function – reference design

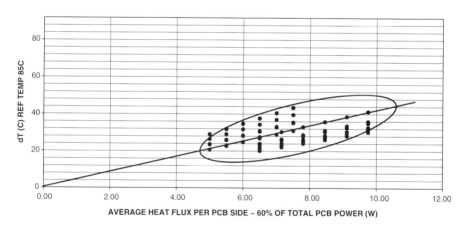

Figure 19.14   Ideal function – optimal design

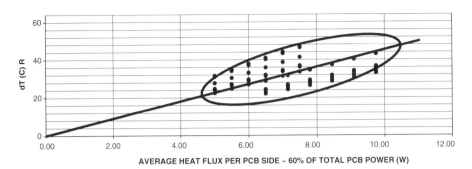

Figure 19.15 Ideal function – tuned design

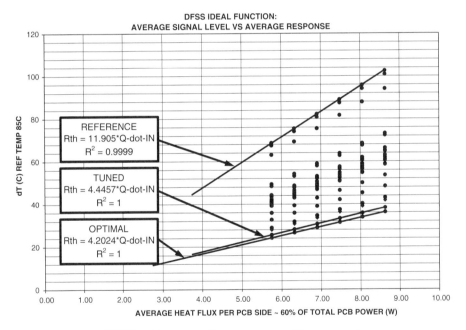

Figure 19.16 Ideal function – average thermal resistance

Table 19.6 Prediction vs confirmation

	REF DESIGN L11	PREDICTION				CONFIRMATION			
		OPTIMAL C1	OPTIMAL GAIN C1-L11	TUNED C2	TUNED GAIN C2-L11	OPTIMAL C1	TUNED C2	TUNED GAIN C2-L11	Delta C2-P Vs C2C
S/N (STB)	−38.79	−29.3	9.51	−31.7	7.11	−29.7	−30.2	8.55	1.4
BETA (Rth)	11.62	3.1	−8.57	5.1	−6.51	4.08	4.31	7.31	0.8
S/N RO = 1 (dT)	−4.05	−3.90	0.15	−3.57	0.48	−3.99	−4.04	0.00	0.5
S/N RO = 1 (Q)	−25.55	−15.90	9.65	−17.96	7.59	−16.4	−17.0	8.56	1.0

Table 19.6 also shows the value of inverting the dT and the heat flux in their roles as response and signal. The dynamic S/N of Q (heat flux) as the response is more effective than dT.

19.4.11 Verification

The next step is a Computation Fluid Dynamic (CFD) simulation. The CFD model will better represent the thermal noise and control factors – in particular the convection and radiation modes of thermal transfer. In a CFD simulation the true ideal function response (heat flux) can be measured.

Besides the initial insight into the control factors, the math model provided insight in the noise factors. Using the Taguchi tolerance design methodology, the noise factors were analyzed for their impact on the variance of dT. Only the "Reference" design at one signal level was evaluated.

First, the correction factor (CF) was calculated per Equation (19.9).

$$CF_{11} = \frac{\left(\sum\limits_{i=1}^{n} y_i\right)^2}{n} \tag{19.9}$$

Then the total sum of squares (ST), total degrees of freedom, and current total variance (VT) were estimated using Equations (19.10), (19.11), and (19.12).

$$S_{T_{11}} = \left(\sum\limits_{i=1}^{n} y_i^2\right) - CF_{11} \tag{19.10}$$

$$df_{T_{11}} = n - 1 \tag{19.11}$$

$$V_{T_{11}} = {}^{S_{T_{11}}}\big/{}_{df_{T_{11}}} \tag{19.12}$$

for $n = 12$ ... CF = 56479.1 ... $S_T = 838.1$ ... $df_T = 11$ and VT = 76.

Next, the level sums of all the noise factors were calculate and summarized in Table 19.7.

Using Equations (19.13), (19.14) and (19.15), the respective sum of squares, degrees of freedom, and variance was calculated for each noise factor (nf).

$$S_{nf} = {}^{\left(nf_1^2 + nf_2^2\right)}\big/{}_{n} - CF_{11} \tag{19.13}$$

$$df_{nf} = (\# levels) - 1 \tag{19.14}$$

$$V_{nf} = {}^{S_{nf}}\big/{}_{df_{nf}} \tag{19.15}$$

Table 19.7 Response table – level sums for all noise factors

		Level 1	Level 2
N	Convection film coefficient of exterior surface	444	379
P	Areas around ECU that air is allowed to flow	418	405
R	External clearance around ECU	416	407
S	Size of PCB	439	384
T	PCB construction	412	411
U	Average % copper per layer	411	412
V	% Thermal power on top side of PCB	434	390
W	Special 1	412	411
X	Orientation of ECU	423	401
Y	Special 2	413	410
Z	PCB - Hsg Contact thermal impedance cm2- K/W	414	410

Using data for factor S (PCB Size), $S_S = (439^2 + 384^2)/2 = 255.5$ and since $df_S = 1$ then V_S also equals 255.5

All the noise factor variances are tabulated in Table 19.8 along with the percent contribution of each noise factor to the total variance. These contributions were calculated use the simple ratio in Equation (19.16).

Table 19.8 ANOVA table for noise factors

		S	DF	V	p %
N	Convection film coefficient of exterior surface	355.2	1	355.2	42%
P	Areas around ECU that air is allowed to flow	15.0	1	15.0	2%
R	External clearance around ECU	6.6	1	6.6	1%
S	Size of PCB	255.5	1	255.5	30%
T	PCB construction	0.1	1	0.1	0%
U	Average % copper per layer	0.0	1	0.0	0%
V	% Thermal power on top side of PCB	162.6	1	162.6	19%
W	Special 1	0.1	1	0.1	0%
X	Orientation of ECU	41.2	1	41.2	5%
Y	Special 2	0.7	1	0.7	0%
Z	PCB - Hsg Contact thermal impedance cm2- K/W	1.2	1	1.2	0%
	TOTAL	838	11	76.2	100%

Table 19.9   Noise levels for CFD N1 and N2

		N1 (BC)	N2 (WC)
N	Convection film coefficient of exterior surface	2	1
P	Areas around ECU that air is allowed to flow	2	1
R	External clearance around ECU	2	1
S	Size of PCB	2	1
V	% Thermal power on top side of PCB	1	2
X	Orientation of ECU	1	2

$$\rho_{nf} = \left(S_{nf} \big/ S_{T_{11}} \right) \times 100 \qquad\qquad (19.16)$$

From this analysis of dT variance due to noise, the noise strategy for the CFD will be changed to a compounding N1 and N2 or "Low Energy" and "High Energy" noise strategy. Using Table 19.7 and Table 19.8 together, it was determined that N1 and N2 should be defined as in Table 19.9.

Generally, the control factors for the CFD DFSS analysis will be the same as the math-model simulation. The technical representation of those factors and their levels however will be tailored to the CFD simulation.

Finally, the ideal function response will be changed to the total thermal energy leaving the subsystem. This is a more direct measurement of the ideal function than the change in temperature.

19.5 Conclusions

Our project demonstrated that an electronic spreadsheet can be an effective and efficient thermal robustness design template (makes better decisions faster). The levels of signal, noise and control factors could quickly and easily modified. Within a day, top-level design decisions could be made about the mechanical configuration with respect to the thermal robustness of an enclosure.

Before this analysis was conducted, we only knew how to tune our designs; now we know which direction we should be going with our designs to make them more thermally robust.

- Type 3 heat sinking features should be employed.
- Material type 2 should be considered as a low cost alternative material type 3.

- Material type 1 should be avoided.
- Exchanging type 1 should be employed.
- Material thickness level 3 should be employed.
- Emissivity level 2 should be employed.

The results represent a potential paradigm shift in our thermal management design strategies. Having experienced the value of the two-step optimization process, we now appreciate the efficiency of focusing on reducing variation before tuning the design for performance or cost. Taguchi robust engineering methodologies support our goal of standardizing our enclosure design decision-making so all product lines are thermally robust.

The knowledge learned from the math-model of the project has had an impact on the approach to next CAE simulation phase. The noise factor strategy was simplified to two levels, levels were refined for the critical control factors, and the response was changed from delta temperature to thermal energy transferred to the environment. The correlation between the CAE results and the math-model will better help us understand the limitations of each when making better decisions faster.

Using the math-model for evaluating thermal robustness in our first DFSS project proved to be a prudent decision. Besides being able to apply our DFSS training sooner, the math-model forces the engineer to understand the subsystem physics and the ideal function.

In conclusion, all goals of this project were met. In fact, one could conclude we exceeded our overall goal of making better decisions faster. By shifting the basis of our decisions to robustness, we will make wiser decisions more effectively.

19.5.1 Acknowledgments

I would like to thank the following individuals:

- Shin Taguchi and Craig Jensen from ASI Consulting Group, LLC, USA for sharing their wisdom, and expertise;
- Eric Devore and Keenen Cluskey of Robert Bosch Corporation, USA for providing me with the opportunity to train and apply the Taguchi Robust Engineering (RE) methodology;
- Mary Albrecht of Albrecht Engineering for sharing her technical expertise in thermal management of automotive electronics;
- John Casari and Eileen Chocallo of Robert Bosch Corporation, USA for editing this chapter.

19.6 Futher Reading

1. ASI Consulting Group LLC (2006) Design for Six Sigma (DFSS) Bosch Project Leader Training Week B, Revision 2.4, Copyright 2006.
2. Rencz, M. and Székely, V. (2004) Structure function evaluation of stacked dies, *20th IEEE SEMI-THERM Symposium, San Jose, CA*

This case study is contributed by Paul Wickett of Robert Bosch Corporation, USA.

20

Pressure Switch Module Normally Open Feasibility Investigation and Supplier Competition

Robert Bosch, LLC, USA

20.1 Executive Summary

A pressure switch module in the transmission control module monitors clutch application (gear selection) state. The baseline design strategy (normally closed) is costly, but meets all usage and performance requirements.

Utilization of the DFSS process provided multiple benefits for improving the pressure switch module. Primarily, a robust assessment was performed to assess the technical feasibility of the normally open switch. Second, using robust engineering, additional performance gains were obtained from the normally open pressure switch design. Finally, the cost benefit potential of

Robust Optimization: World's Best Practices for Developing Winning Vehicles,
First Edition. Subir Chowdhury and Shin Taguchi.
© 2016 Subir Chowdhury, Shin Taguchi, and ASI Consulting Group, LLC.
Published 2016 by John Wiley & Sons, Ltd.

the new design was maintained by performing a robust assessment between two separate suppliers using both technical and commercial criteria.

The noise strategy was selected to ensure a common test environment between suppliers, while the control factors were independently developed.

One supplier's optimal design did not confirm, however the achieved S/N was equal to the current production design for no dB gain. The second supplier marginally confirmed a 10.5 dB gain over current production.

20.2 Introduction

Modern step transmissions use electronic monitoring of its critical aspects to maintain proper performance throughout its life. One such monitoring system measures the state of pressure in the clutches. This information is used by the system for calibration, gear selection, and shift quality. A mechatronic pressure switch that converts clutch pressure to electrical signal performs the measurement.

The input energy to the switch from the transmission is the presence of pressure above a set level in the particular clutch. During the shift event, the pressure ramps from zero to full pressure as the clutch engages and then back again as the clutch disengages.

The switch indicates the clutch state by changing the voltage across it in response to the pressure changes. The transmission control unit (TCU) uses this signal to trigger the next action in the shift event. This will apply the clutches at the appropriate time to maintain shift quality, and proper gear selection.

The switch must measure millions of pressure cycles while exposed to contamination, a large temperature range, and the aging effects of transmission fluid. The switch must clearly indicate an actuated state before the clutch pressure exceeds a transition value. Similarly, the switch must change state prior to a second transition point, to indicate when the clutch pressure is properly exhausted.

20.2.1 Current Production Pressure Switch Module – Detailed

The pressure switch module (Figure 20.1) contains:

- a backing plate to maintain flatness and provide support;
- a leadframe to connect the switching element to the transmission control unit;
- elastomer seals to mate and seal to the transmission valve body;
- switching elements to detect the clutch pressure;
- bushings to maintain proper bolt clamp load.

Figure 20.1 Pressure switch module

20.2.2 Current Production (N.C.) Switching Element – Detailed

Each switching element (Figure 20.2) contains:

- a lower contact which is circuit ground;
- a diaphragm to isolate the switch internals from the direct acting fluid;
- a piston to transfer the motion of the diaphragm to the disc;
- a spring disc that changes state in reaction to the piston force;
- an insulator (lead) which is circuit power;
- elastomer seals to mate and seal to the transmission valve body.

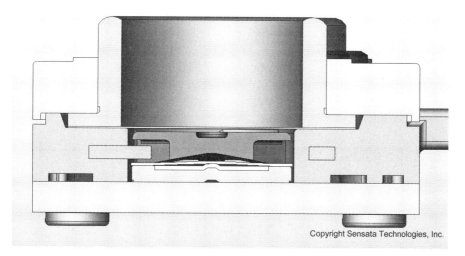

Copyright Sensata Technologies, Inc.

Figure 20.2 Pressure switch element

20.3 Objective

- An objective of this project was to investigate the feasibility of changing the pressure switch module from a normally closed design to a normally open design. The target was to prove the normally open design was as robust as, or more so, than the current production normally closed design. This feasibility was verified by conducting a robust assessment between the normally closed and normally open switch design.
- Next, we optimized each suppliers design using robust engineering methodology. Close collaboration with the suppliers established control factors and individual inner arrays to fit their unique designs.
- Cost reduction was the main driver for investigating this change. Accordingly, another objective was to maximize the cost benefit from changing from normally closed to normally open technology. This was accomplished by running a robust assessment of the optimized designs from each supplier. The criteria used in this final robust assessment included both technical and commercial criteria.
- An additional objective for the supplier competition was to locate a second source for the PSM. The current production supplier, since product inception, has lost the exclusivity of market share that led to their initial sourcing. The proposed new source would be utilizing a technology outside of their established expertise. Robust engineering will speed up the development of the normally open switch at the prospective supplier.
- Another objective was to develop a noise strategy to use in future development of PSM technology. A 20% reduction in test time, equipment and/or man-hours was the target.
- A broader objective of this study is to establish a precedent to use robust engineering to reduce validation testing, time, and cost to source a new supplier. This methodology can be adapted to all components, materials, and processes. This effort supports our mission to improve the quality and cost of our core technologies.

20.4 Robust Assessment

20.4.1 Scope and P-Diagram

The scope of the ideal function is the switching element inside of the pressure switch module as shown in Figure 20.3. From this scope, two detailed P-diagrams were created and are partially shown in Figure 20.4. These P-diagrams were the basis for the noise, signal and control factors strategies.

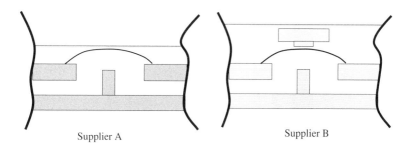

Supplier A

Supplier B

Figure 20.3 Scope

20.4.2 Ideal Function

The scope of the ideal function for the PSM investigation was narrowed to the pressure point at which the switch actuated and released. This differs from the system level application of the pressure signal, because the test stand does not readily provide this measurement. In addition, the customer controls the use of the signal.

During the actuation and release cycles, the switch is required to change state at set pressure values. There are separate limits for actuation and release pressures, to maintain system performance. Accordingly, a nominal-the-best ideal function with indicative factors was used to analyze the actuation and release pressures of the switch. This will not penalize the two states for being

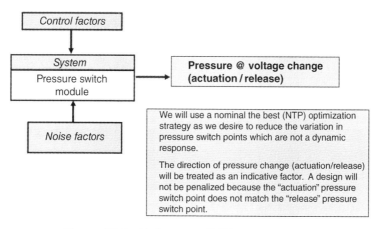

Figure 20.4 P-diagram (DFSS investigation)

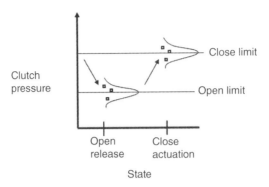

Figure 20.5   Ideal function

apart from each other, but will take into account the variability around each point. See Figure 20.5.

20.4.3 Noise Strategy

The noise factors for the pressure switch module were identified through collaboration between the suppliers, Bosch, and the end customer. A list, as seen in Table 20.1, shows the resulting noise factors. The two main noise factors were contamination robustness and material aging.

20.4.4 Testing Criteria

Test equipment limitations forced the predominant noise characteristics to be separated into two different testing procedures.

The first test was a continuously monitored contamination test. The testing criteria for this test were a sum of the worst-case requirements from the customer. Contamination in the oil was 90 mg/L, with a 4:1:1 mixture of Synform, ECCC, and Carbonyl Iron. The test parts were mounted horizontally, submerged in swirling oil. The oil temperature was maintained at 135 °C, with the supply pressure going to the parts set at 300 psi (machine maximum). Testing of the part was carried out as shown in Figure 20.6.

The pass criterion was a steady and correct signal at each of the dwell points. Only the state change was critical for this test. As such, the switch state during the pressure ramp was not monitored.

The contamination testing was not directly included into the signal-to-noise calculations. It was, however treated as attribute data, where a failure for

Table 20.1   Noise factors

Noise Factors (Brainstormed)
Material aging
Contamination (In the fluid)
(1) Conductive
(2) Nonconductive
Temperature profile (transmission)
Oil properties (ex. viscosity, etc)
Misuse of component
Improper setup of trans (ex. Incorrect fluid)
PSM to VB interface
Assembly of module to the valve body
(1) Welding parameters
Oscillation of supplied pressure
Controlled current/voltage/etc.
Submerged or not?
Horizontal or vertical orientation

contamination would penalize the S/N for that design for the life cycle test by setting it to the lowest passing value minus 3 dB.

The second test was a life cycle test. The pressure profile was a 5 Hz square wave from 0 to 300 psi (5 switch actuation/release cycles per second). The oil temperature was maintained at 135 °C, with no contamination added. The target life was 2 million cycles. When failures occurred, the pressure cycling was discontinued for that sample.

Throughout the course of the test, the pressure cycling was stopped to perform a manual "pressure-to-switch" measurement. This involved a manual ramp of the pressure of each part from maximum to zero, and then back up again. The pressure at which the switch changed state was recorded as either actuation (increasing pressure) or release (decreasing pressure).

20.4.5   Control Factors and Levels

A standard orthogonal L-18 matrix was used for each suppliers design concept. This provided eight control factors. Though the selection of the control factors started with the P-diagram, the final set of factors was

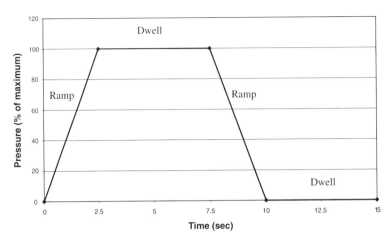

Figure 20.6 Contamination test pressure cycle

determined through teamwork between Bosch and the individual supplier. See Table 20.2 and Table 20.3 for details. Note, that for confidentiality, each supplier was not provided the control factors from the other.

20.4.6 Test Data

The results, given in Table 20.4, Table 20.5, Figure 20.7, and Figure 20.8 are a compilation of the acquired data. This consists of the actuation and release pressure points at 0 and 2M cycles.

Table 20.2 Control factors – supplier A

ID	Control factor	Level 1	Level 2	Level 3
A	Heat stake presence	Nominal	Not present	
B	Vent size	Nominal	Nominal – 1.5Tol	Nominal +1.5Tol
C	Kapton surface depth	Nominal	Nominal – 1.5Tol	Nominal +1.5Tol
D	Disc seat depth	Nominal	Nominal – 1.5Tol	Nominal +1.5Tol
E	Lower contact depth	Nominal	Nominal – 1.5Tol	Nominal +1.5Tol
F	Overtravel depth	Nominal	Nominal – 1.5Tol	Nominal +1.5Tol
G	Disc actuation pressure	Nominal	Nominal – 1.5Tol	Nominal +1.5Tol
H	Disc differential pressure	Nominal	Nominal – 1.5Tol	Nominal +1.5Tol

Note: Levels 2 and 3 set by shifting nominal values beyond the tolerance limit.

Table 20.3 Control factors – supplier B

ID	Control factor	Level 1	Level 2	Level 3
A	Rivet pressure	Level 1	Level 2	–
B	Outer contact to diaphragm distance	Nominal	Nominal – 1.5Tol	Nominal +1.5Tol
C	Center point to outer point distance	Nominal	Nominal – 1.5Tol	Nominal +1.5Tol
D	Disc height	Nominal	Nominal – 1.5Tol	Nominal +1.5Tol
E	Plating thickness	Nominal	Nominal – 1.5Tol	Nominal +1.5Tol
F	Piston thickness	Nominal	Nominal – 1.5Tol	Nominal +1.5Tol
G	Filter size	Nominal	Nominal – 1.5Tol	Nominal +1.5Tol
H	Seal hardness	Nominal	Nominal – 10	Nominal + 10

Note: Levels 2 and 3 set by shifting nominal values beyond the tolerance limit.

20.4.7 Data Analysis

There are separate limits for actuation and release pressures, to maintain system performance. Accordingly, a nominal-the-best ideal function with indicative factors was used to analyze the actuation and release pressures of the switch. This will not penalize the two states for being apart from each other, but will take into account the variability around each point due to aging noise.

The NTB using indicative factors equation for signal-to-noise is:

$$S/N = 10 * \log \left(\frac{\left[Avg_{Act} + Avg_{Rel} \right]^2}{\left[\dfrac{\delta_{Act}^2 + \delta_{Rel}^2}{2} \right]} \right)$$

The data analysis calculations for each supplier are listed in Table 20.6 and Table 20.7. A large spread of S/N ratio for both suppliers indicates that the noise strategy was effective. Also worth noting is that two runs for Supplier A were infeasible and were penalized to the lowest S/N ratio minus 3 dB. These designs were infeasible because the combination of control factors led to a geometric condition that increased the amount of stress, leading to premature failure.

Table 20.4 Test data – supplier A

Exp	A	B	C	D	E	F	G	H	N1 - 0 cycles Actuation S1	S2	S3	Release S1	S2	S3	N2 - 2M cycles Actuation S1	S2	S3	Release S1	S2	S3
1	1	1	1	1	1	1	1	1	137.59	137.81	134.05	104.80	104.94	102.09	112.27	108.43	107.75	83.63	75.57	77.26
2	1	1	2	2	2	2	2	2	1291.24	1413.77	1345.23	1181.33	1372.32	1295.54	0.00	0.00	0.00	0.00	0.00	0.00
3	1	1	3	3	3	3	3	3	270.79	257.41	259.82	261.35	272.10	227.95	278.27	194.81	166.68	272.63	186.36	158.08
4	1	2	1	1	2	2	3	3	651.18	717.72	808.13	560.41	584.05	701.22	0.00	0.00	0.00	0.00	0.00	0.00
5	1	2	2	2	3	3	1	1	169.37	180.64	181.76	156.89	168.79	168.94	140.86	140.77	145.74	132.66	136.82	141.68
6	1	2	3	3	1	1	2	2	126.15	136.72	143.16	100.65	124.16	128.20	120.86	105.19	137.67	94.93	86.74	86.88
7	1	3	1	2	1	3	2	3	100.30	94.69	98.93	50.09	42.46	46.70	89.64	83.69	87.73	48.26	42.17	45.06
8	1	3	2	3	2	1	3	1	200.12	201.48	190.60	178.54	177.06	170.16	166.77	179.29	167.82	156.29	173.95	161.33
9	1	3	3	1	3	2	1	2	282.68	314.14	295.67	245.26	264.18	258.78	210.46	210.79	212.45	199.66	207.25	208.81
10	2	1	1	3	3	2	2	1	312.40	312.34	323.21	275.81	277.02	292.79	259.67	237.15	152.77	256.02	236.54	150.63
11	2	1	2	1	1	3	3	2	173.84	162.84	160.84	148.32	132.56	131.40	154.10	146.56	152.11	135.44	134.52	131.04
12	2	1	3	2	2	1	1	3	115.15	119.28	124.63	91.02	95.30	96.33	118.04	115.91	118.03	98.49	98.29	99.09
13	2	2	1	2	3	1	3	2	366.32	234.70	230.91	225.84	218.07	216.64	233.20	221.64	214.71	229.24	216.80	208.38
14	2	2	2	3	1	2	1	3	151.49	157.26	147.23	108.63	115.86	110.72	132.49	135.77	118.71	101.08	102.88	97.65
15	2	2	3	1	2	3	2	1	110.44	101.69	109.10	89.63	81.12	91.80	102.10	97.08	92.77	89.68	84.97	82.68
16	2	3	1	3	2	3	1	2	165.43	159.62	157.79	148.48	142.87	129.50	131.89	122.57	134.42	119.47	116.61	127.88
17	2	3	2	1	3	1	2	3	143.00	112.22	133.44	134.21	116.81	138.56	120.02	116.81	143.11	116.84	112.57	136.82
18	2	3	3	2	1	2	3	1	160.50	161.44		131.77	140.44		150.82	149.52	148.51	124.00	133.66	135.55

Factors: A Heat Stake Presence, B Vent Size, C Kapton Surface Depth, D Disc Seat Depth, E Lower Contact Depth, F Overtravel Depth, G Disc Actuation Pressure, H Disc Differential Pressure

Table 20.5 Test data – supplier B

Exp	A	B	C	D	E	F	G	H	N1 - 0 cycles Actuation S1	S2	S3	Release S1	S2	S3	N2 - 2M cycles Actuation S1	S2	S3	Release S1	S2	S3
1	1	1	1	1	1	1	1	1	155.36	151.24	194.39	137.43	121.44	172.38	5.58	164.23	9.45	0.71	143.47	5.98
2	1	1	2	2	2	2	2	2	104.94	90.31	120.48	77.69	63.42	83.66	137.70	131.14	137.16	123.33	119.93	109.45
3	1	1	3	3	3	3	3	3	242.09	163.66	202.03	112.99	147.39	187.24	0.00	187.31	203.11	0.00	165.57	187.17
4	1	2	1	1	2	2	3	3	145.37	167.21	140.65	126.56	134.07	123.26	176.90	178.55	186.19	152.79	162.70	170.72
5	1	2	2	2	3	3	1	1		105.34	104.67		94.19	82.25	120.08	138.37	120.66	111.82	129.35	103.58
6	1	2	3	3	1	1	2	2	139.08	189.54	178.44	98.40	127.93	150.68	161.06	177.20	182.48	140.64	140.86	144.37
7	1	3	1	2	1	3	2	3	152.71	159.60	164.78	138.26	140.02	127.83	161.10	157.56	151.47	156.66	146.53	137.57
8	1	3	2	3	2	1	3	1	196.02	179.38	182.09	168.99	138.06	141.32	219.48	194.48	199.97	188.16	168.34	175.78
9	1	3	3	1	3	2	1	2		184.10	190.41		129.93	154.65	164.73	0.00	197.11	142.49	0.00	171.59
10	2	1	1	3	3	2	2	1	273.46	252.97	218.06	256.20	219.14	177.09	216.17	211.58	197.41	130.44	183.24	172.23
11	2	1	2	1	1	3	3	2	201.47	190.10	192.80	169.34	167.01	143.32	162.46	175.05	178.97	130.44	144.04	145.25
12	2	1	3	2	2	1	1	3	146.12	122.41	151.53	145.07	108.28	129.53	144.57	151.24	153.98	134.21	142.20	145.41
13	2	2	1	2	3	1	3	2	147.32	161.24	122.48	111.38	142.73	86.25	168.19	146.41	115.43	141.70	124.38	88.36
14	2	2	2	3	1	2	1	3	211.39	199.24	256.89	181.93	177.58	227.89	183.81	201.30	237.06	169.78	193.72	218.17
15	2	2	3	1	2	3	2	1	151.95	136.34	172.71	119.36	124.40	150.05	169.19	141.27	161.96	156.54	135.75	151.60
16	2	3	1	3	2	3	1	2	187.96	173.57	235.42	166.56	146.27	211.65	183.16	182.08	227.60	164.10	166.62	205.14
17	2	3	2	1	3	1	2	3	182.24	184.47	132.09	162.22	163.16	71.49	188.07	176.23	114.41	174.81	167.92	75.68
18	2	3	3	2	1	2	3	1	190.54	201.89	140.37	170.77	153.24	69.18	0.00	0.00	131.21	0.00	0.00	94.73

Factors: A Rivet Pressure, B Outer Contact to Diaphragm Distance, C Center Point to Outer Point Distance, D Disc Height, E Plating Thickness, F Piston Thickness, G Filter Size, H Seal Hardness

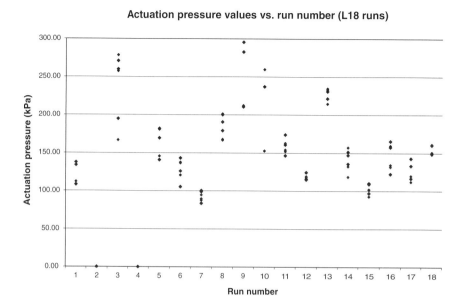

Figure 20.7 Example of response data – supplier A

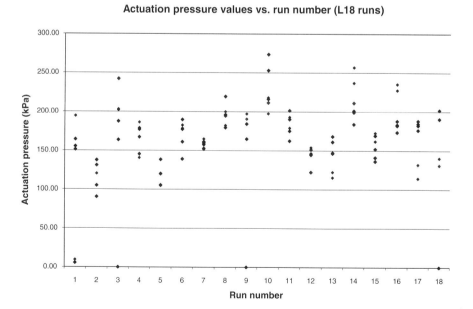

Figure 20.8 Example of response data – supplier B

Table 20.6 Data analysis – supplier A

| | | | | | | | | | | | | | | SN Range |
| | | | | | | | | | | | | | | 21.57 |

Experiment Number	Heat Stake Presence	Vent Size	Kapton Surface Depth	Disc Seat Depth	Lower Contact Depth	Overtravel Depth	Disc Actuation Pressure	Disc Differential Pressure				Indicative Factors			
	A	B	C	D	E	F	G	H	Avg Act	Avg Rel	Sigma Act	Sigma Rel	Avg Sigma	Sum Squared Avg	S/N Ratio
1	1	1	1	1	1	1	1	1	122.98	91.38	14.93	14.06	210.2364	45952.067405	23.40
2	1	1	2	2	2	2	2	2	165.25	143.97					15.00
3	1	1	3	3	3	3	3	3	237.96	229.74	45.83	48.28	2215.549	218748.954403	19.94
4	1	2	1	1	2	2	3	3	165.25	143.97					15.00
5	1	2	2	2	3	3	1	1	159.85	150.96	19.63	16.11	322.3998	96607.000278	24.77
6	1	2	3	3	1	1	2	2	128.29	103.59	13.95	18.31	264.9391	53771.116996	23.07
7	1	3	1	2	1	3	2	3	92.50	45.79	6.57	3.17	26.5616	19123.524848	28.57
8	1	3	2	3	2	1	3	1	184.35	169.55	15.42	8.95	158.9287	125244.384235	28.97
9	1	3	3	1	3	2	1	2	254.36	230.66	48.30	28.69	1577.995	235244.077053	21.73
10	2	1	1	3	3	2	2	1	266.26	248.13	65.22	51.57	3456.414	264598.272345	18.84
11	2	1	2	1	1	3	3	2	158.38	135.55	9.62	6.49	67.31069	86395.040853	31.08
12	2	1	3	2	2	1	1	3	118.51	96.42	3.36	3.01	10.18293	46193.830256	36.57
13	2	2	1	2	3	1	3	2	250.24	219.16	57.37	7.43	1673.511	220343.088118	21.19
14	2	2	2	3	1	2	1	1	140.49	106.14	14.20	6.78	123.8392	60826.110270	26.91
15	2	2	3	1	2	3	2	1	102.20	86.65	6.79	4.33	32.46341	35662.559922	30.41
16	2	3	1	3	2	3	1	2	145.29	130.80	17.78	12.64	237.9834	76224.675773	25.06
17	2	3	2	1	3	1	2	3	128.10	125.97	13.57	11.75	161.1605	64549.786422	26.03
18	2	3	3	2	1	2	3	1	154.16	133.08	6.28	6.02	37.84352	82507.851667	33.39
									165.25	143.97					25.00

The first step in the Taguchi two-step optimization is to focus on variation. The response plots for all data are shown in Figure 20.9, Figure 20.10, Figure 20.11, Figure 20.12, Figure 20.13, and Figure 20.14. The customer had no particular targets for the actuation and release points. Their only requirement is that they are within tolerance. Therefore, reducing variation was the primary goal.

From Figure 20.9, the NTB S/N analysis indicates that, for Supplier A, factor E (Lower Contact Depth) had the largest impact on robustness across all noise levels with a range of 5.65 db. Three additional factors had S/N ranges near 5 dB. In order of magnitude, the factors are Factor C (Kapton Surface Depth), Factor A (Heat Stake Presence), and Factor F (Overtravel Depth). The dB ranges for these variables were 5.51, 5.45, and 4.83 dB, respectively.

From Figure 20.12, the NTB S/N analysis indicates that, for Supplier B, factor E (Plating Thickness) had the largest impact on robustness with a range of 6.50 db. Two additional factors had S/N ranges near 5 dB. These factors are Factor F (Piston Thickness) and Factor C (Center Point to Outer Point Distance) with dB ranges of 5.45 and 5.23 dB, respectively.

Table 20.7 Data analysis – supplier B

	A Rivet Pressure	B Outer Contact to Diaphragm Distance	C Center Point to Outer Point Distance	D Disc Height	E Plating Thickness	F Piston Thickness	G Filter Size	H Seal Hardness	Avg Act	Avg Rel	Sigma Act	Sigma Rel	Avg Sigma	Sum Squared Avg	S/N Ratio
														SN Range	24.36
1	1	1	1	1	1	1	1	1	113.37	96.90	83.39	74.34	6239.857	44215.155076	8.50
2	1	1	2	2	2	2	2	2	120.29	96.24	19.18	24.69	488.9032	46886.467911	19.82
3	1	1	3	3	3	3	3	3	166.37	133.39	85.42	71.04	6171.188	89855.757840	11.63
4	1	2	1	1	2	2	3	3	165.81	145.01	18.73	19.84	372.0922	96613.423929	24.14
5	1	2	2	2	3	3	1	1	117.82	104.24	13.82	17.84	254.6331	49310.110657	22.87
6	1	2	3	3	1	1	2	2	171.30	133.81	18.37	18.87	346.7666	93093.332544	24.29
7	1	3	1	2	1	3	2	3	157.87	141.15	5.08	9.69	59.79636	89410.269240	31.75
8	1	3	2	3	2	1	3	1	195.24	163.44	14.39	19.76	298.7087	128649.668565	26.34
9	1	3	3	1	3	2	1	2	147.27	119.73	83.21	68.68	5820.176	71290.281606	10.88
10	2	1	1	3	3	2	2	1	228.27	198.72	28.75	32.66	946.8687	182321.029420	22.85
11	2	1	2	1	1	3	3	2	183.47	149.90	14.05	15.16	213.6383	111136.557012	27.16
12	2	1	3	2	2	1	1	3	144.97	134.12	11.61	14.15	167.5433	77890.576891	26.67
13	2	2	1	2	3	1	3	2	143.51	115.80	20.86	24.96	529.2111	67243.577720	21.04
14	2	2	2	3	1	2	1	3	214.95	194.84	27.07	23.37	639.357	167929.073472	24.19
15	2	2	3	1	2	3	2	1	155.57	140.96	14.88	16.85	252.6552	87932.512001	25.42
16	2	3	1	3	2	3	1	2	198.30	176.72	26.26	25.76	676.5768	140641.375477	23.18
17	2	3	2	1	3	1	2	3	162.92	135.88	31.47	48.48	1670.245	89281.639200	17.28
18	2	3	3	2	1	2	3	1	110.67	81.32	90.00	73.11	6722.526	36858.560201	7.39
									161.00	136.79					20.86

Indicative Factors · Experiment Number

20.4.8 Prediction and Confirmation

The suppliers had individual tuned designs that differed from their maximum S/N design. The tuning was based on reducing cost and increasing manufacturability. Adjustments to the means were not necessary; however, the mean predictions were used to double check that they were within specification limits.

Table 20.8 summarizes prediction and confirmation. As indicated by Table 20.9, above, the parts provided from Supplier A outperformed those of Supplier B. The average pressures for both suppliers were within the specification; however, variation outside of specification was present between 0 and 2M cycles. Supplier A exhibited a marginal confirmation (19.62 dB gain predicted vs. 10.49 dB gain confirmed). Supplier B, however, did not confirm (17.55 dB gain predicted vs. −0.38 dB gain confirmed). The lack of confirmation may be attributed to the supplier changing other design parameters to address concerns with other requirements not included in this investigation.

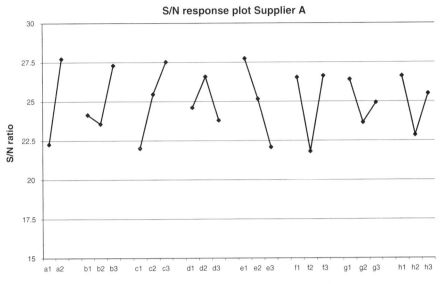

Figure 20.9   S/N response – supplier A

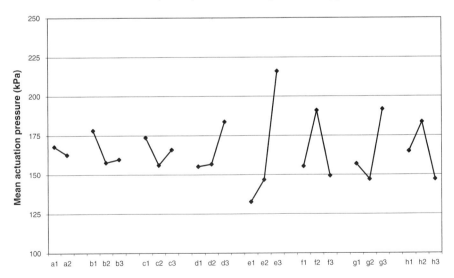

Figure 20.10   Actuation pressure response – supplier A

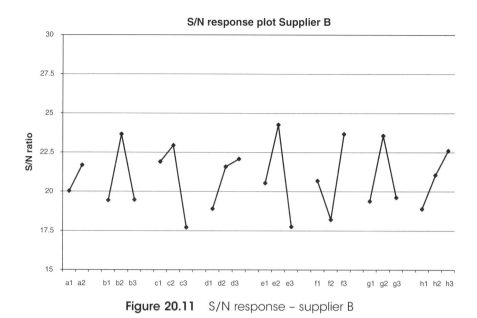

Figure 20.11   S/N response – supplier B

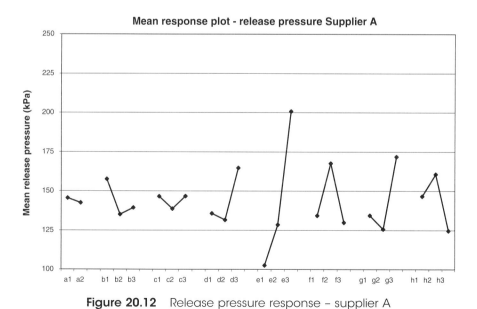

Figure 20.12   Release pressure response – supplier A

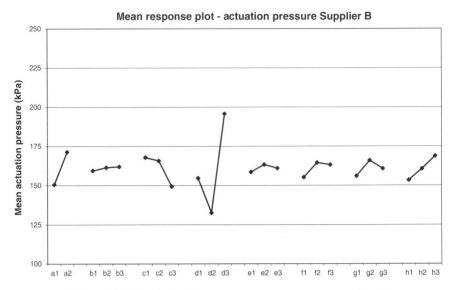

Figure 20.13 Actuation pressure response – supplier B

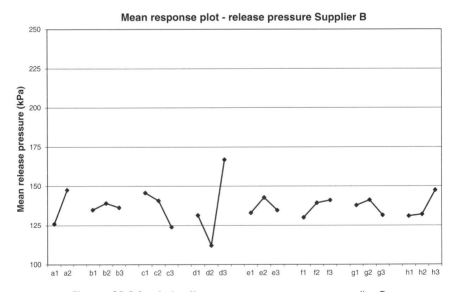

Figure 20.14 Actuation pressure response – supplier B

Table 20.8 Prediction and confirmation (summary)

	Prediction			Confirmation		
	Mean act	Mean rel	S/N ratio	Mean act	Mean rel	S/N ratio
Baseline (current production)	142.31	124.97	20.45	142.31	124.97	20.45
Supplier A tuned design	127.22	103.16	40.07	150.21	128.61	30.93
Supplier A gain			*19.62*			*10.49*
Supplier B tuned design	165.66	154.42	38.00	145.85	119.64	20.06
Supplier B gain			*17.55*			*−0.38*

20.4.9 Verification

The next step is to run verification on a complete pressure switch module (4 switches per module). This will be completed during the design validation phase, which is in its preliminary phase at this point. Verification will include all customer specification requirements.

20.5 Summary and Conclusions

Our project demonstrated that the normally open PSM is a technically feasible design. It was proven that the normally open design was capable of being more robust than the normally closed production PSM design.

After optimization, each supplier exhibited gains over their initial normally open designs. Supplier A's tuned design was a 7.5 dB gain over their individual baseline, and Supplier B's tuned design was 11.5 dB higher than their baseline.

The normally open PSM design is less expensive than the normally closed. With each supplier's design proving technically capable, they were more confident to improve their commercial approach.

Another result of this investigation, Supplier B, a newcomer to transmission pressure sensing, was proven to make a switch that meets the requirements. Because of their failure to confirm, they need to do additional work to achieve the level of robustness of Supplier A. However, their design is as robust as the current production switch, at a lower cost.

Good separation of the individual design combinations was shown at the 1M cycle mark. Response plot relative levels were the same using between 1M

Table 20.9 Prediction and confirmation (raw data)

	N1 - 0 cycles					N2 - 2M cycles							
Supplier A Optimum design	157.70	155.99	161.89	132.20	128.57	137.83	141.35	145.59	138.75	121.92	127.63	123.49
Supplier B Optimum design	156.57	151.08	107.04	94.31	126.55	67.94	148.71	152.13	159.56	136.85	147.52	144.65

and 2M cycles. For future development, 1M cycles will be used to reduce the amount of time on the test stand by 50%.

The gains achieved through this investigation show how robust engineering can be applied to other aspects of the transmission control module manufacture. Similar design competitions can be preformed on all supplied components, to obtain the most robust product at the lowest cost.

The effort of our development was decreased due to the genuine cooperation between our customer, the suppliers, and ourselves by using robust engineering as the common nomenclature, methods and measures.

20.5.1 Acknowledgments

I would like to thank the following individuals:

- Craig Jensen of ASI Consulting Group, LLC, USA for sharing his wisdom, and expertise;
- Len Buckman of Robert Bosch, LLC, USA for providing me with the opportunity to train and apply the Taguchi Robust Engineering (RE) methodology;
- John Casari and Karl Gage of Robert Bosch, LLC, USA for assistance with editing this case study.

This case study is contributed by Andrew Yermak of Robert Bosch, LLC, USA.

Part Four
Manufacturing Process Optimization

21

Robust Optimization of a Lead-Free Reflow Soldering Process

Delphi Delco Electronics Systems, USA and ASI Consulting Group, LLC, USA

21.1 Executive Summary

This case study will focus on Dr. Taguchi's Robust Engineering methodology, measurement methods and experimental results for the optimization of a lead-free surface mount (SMT) reflow process for use in an automotive electronics application. Traditional optimization approaches focus on maximizing the response variable while the robust approach focuses on consistent results regardless of variation in noise factors.

The Robust method was utilized in the development of a lead-free process for manufacturing an Automotive SMT product. Major factors that can create variation in a lead-free process were identified, including lead-free solder paste brand, paste print speed, oven reflow temperatures and times, and reflow environment. Several noise factors were studied including volume of solder paste, location of components on the board, and lead frame plating materials, namely tin and palladium/nickel/gold. A series of measurements were made

Robust Optimization: World's Best Practices for Developing Winning Vehicles,
First Edition. Subir Chowdhury and Shin Taguchi.
© 2016 Subir Chowdhury, Shin Taguchi, and ASI Consulting Group, LLC.
Published 2016 by John Wiley & Sons, Ltd.

on the lead-free product that assessed the strength and reliability of lead-free solder joints, measurements such as visual scoring, cross-section, surface insulation resistance, and pull strength. Using the Robust experimental design, these measurements were optimized to create high quality and reliable lead-free SMT solder joints that were the most insensitive to the noise. In essence, quality was increased by using variable measurements rather than by counting attributes (good/bad). Overall, a gain of 2.1 dB was realized. In Robust terms, this equates to reducing variation in the lead-free process by ~22%. This study also revealed which of the processing factors were most significant in controlling the lead-free process. The results of this study and the use of Robust Engineering methodology provide a means for developing a full range of lead-free technology, components and products used on Automotive Electronics.

21.2 Introduction

One of the challenging new fields in the electronics industry is the implementation of lead free into electronic products. This change is being driven primarily via legislation, especially in Europe, specifically through the WEEE and RoHS Directive in Europe, in its current revision banning the use of lead in electronics by 2006 [1,2].

An essential element to developing these lead-free products is the implementation of the lead-free manufacturing process. The lead-free manufacturing process poses new challenges not encountered in the traditional tin/lead world of electronics. One key difference in the lead-free manufacturing footprint from traditional processes is the reduced wetting potential of the lead-free solders that will be used. The physics of wetting dictate this reduced wetting since the lead-free alloys have higher surface tensions than tin/lead systems, therefore making it more difficult to solder parts [3].

In addition to the wetting challenges, lead-free alloys require higher melting temperatures than their tin/lead counterparts, increasing the stresses on the components to be soldered. An important element to lead-free manufacturing is limiting the maximum process temperatures and thermal energies to which components are exposed. For example, in the recently released IPC/JEDEC J-STD-020B, a component classification industry standard, lead-free packaged parts are only required to be classified up to peak oven temperatures of 245 °C or 250 °C depending upon package size. Once a process is put into place that can produce soldered parts within such temperature requirements, only then can a sufficient component supply base be made available that can match product requirements.

For the automotive electronics industry, the reliability of the manufactured electronic product is important due to the stressful environments seen by these products. In this high-volume business, first time quality (FTQ) must be high, and the need for an effectively optimized manufacturing process is an absolute requirement. This is especially true in the emerging field of lead free.

This chapter discusses the optimization of a surface mount only (SMT) process for a lead-free automotive electronics assembly process. SMT was chosen due to the drop-in replacement potential for this soldering technique, as opposed to wave soldering or selective soldering, which are much more immature fields in terms of lead-free development [4]. This immaturity is partially the result of the huge capital investments that are necessary for wave soldering or selective soldering since dedicated soldering pots are required because Pb poisoning is problematic [5]. The Robust Engineering methodology was used to optimize this lead-free SMT manufacturing process. This methodology is a powerful tool for selecting the best manufacturing process parameters that mitigate manufacturing variation. This process will be discussed, as will the specifics of the lead-free investigation.

21.3 Experimental

21.3.1 Robust Engineering Methodology

To evaluate the SMT lead-free process for an automotive electronics assembly process, the Robust Engineering approach was used. In general, the Robust Engineering strategy is to find the combination of factors that produce the most consistent response, across various noise conditions. In other words, the goal is to achieve consistent response regardless of variations in manufacturing, environment, and customer's usage conditions.

The first step was to determine what response to measure. It was decided to optimize solder joint wetting and then verify strength of the optimum combination using thermal cycling. Pull strength was also evaluated; however, the repeatability of this response is suspect. Solder wetting was evaluated using visual scoring (numerical grading scale).

The second step was to identify the key process factors that effect wetting results. This was done by developing a Process Map for the SMT process (PMAP) and a Failure Mode and Effects Analysis table (FMEA). The PMAP is produced by "walking" the process and developing a detailed map of each stage of the manufacturing process. The inputs and desired outputs of each stage are also recorded. This format identifies inputs as either Noise or Control factors, which is critical to a Robust Engineering approach. Critical

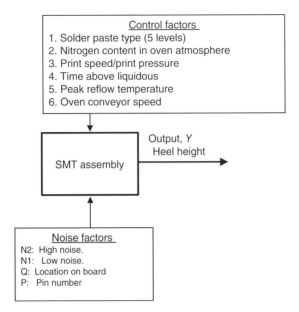

Figure 21.1   Parameter diagram showing critical control and noise factors

inputs to the proper output conditions can be tagged. The critical factors from the PMAP are listed on the Parameter Diagram which is shown in Figure 21.1.

The FMEA tabulates potential failure mechanisms at each stage of the process, and the causes of those failures. The level of severity of these failures to producing the proper output steps allows for the ranking of importance of process inputs.

From the PMAP and FMEA, the following SMT process inputs were selected for testing for their effects on lead-free solder joints. See Table 21.1.

Table 21.1   Lead-free process inputs for evaluation

Pb free process control factors
Paste brand
Paste print speed
Oven peak temperature
Time above liquidous
Oven conveyor speed
Oven nitrogen level

Table 21.2 Lead-free process noise factors

Noise factor	Low noise	High noise
Thickness of paste brick	Thick stencil	Thin stencil
IC plating metalization	Sn plating on Cu	Ni/Pd/Au plating on Cu
Paste dry out time	Reflow immediately	Reflow after 1/2 hour under fan
Pin to pin location	Multiple pins tested	Multiple pins tested
Component location on board	Multiple board locations of same part tested	Multiple board locations of same part tested

The next main task in the Robust Engineering approach is the selection of noise factors. Noise factors are defined as factors that affect the output of the process but either can't be controlled, or are too expensive to control. Since wetting grade was used as the measure for process capability, key noise factors that effect wetting were selected. In the Robust approach, test cells are built at each of two noise levels, low and high. All factors that contribute favorably to the output are considered low noise, while detrimental inputs are considered high noise. Several noise factors were identified via the PMAP. Those that were suspected to have a large effect on the output of wetting and that can often be seen in production settings were selected. Table 21.2 shows the factors that were selected, as well as the manner by which these noise factors were implemented during the experimental runs.

A key strategy of the Robust approach is to find a combination of control factors which result in the response factor being insensitive to noise factors. For this lead-free process optimization, this strategy is intended to reduce variation in wetting results over time. To study the effect of control factors, a special orthogonal array, known as an L18, was used. The L18 has the special property of evenly spreading interaction affects across all control factors. This minimizes the risk of confounding interactions while also minimizing the number of combinations to test.

For each of the 18 combinations, boards are assembled at each noise condition. The response variable is measured for each noise condition and each run. Using these results, the Signal to Noise (S/N) ratio is calculated for each of the 18 runs. The equation for the Nominal the Best S/N ratio is: $10 * Log(\bar{y}^2/\sigma^2)$, where $\bar{y}$ is the average of the responses for that run and σ is the standard deviation of the responses for that run.

A higher S/N ratio indicates that the combination has more consistent results across noise conditions. S/N is therefore an index of "robustness." Dr. Genichi Taguchi defines robustness as "the state where the technology, product, or process performance is minimally sensitive to factors causing variability (either in the manufacturing or user's environment) at the lowest unit cost" [6].

The next step is to use the results of this experiment to predict the optimum combination, which is likely to be something different from one of the 18 original runs. The signal to noise is also predicted for this combination and a baseline combination. The difference between the baseline S/N and optimum S/N is the S/N Gain.

The final step is confirmation of the signal to noise for the baseline, optimum and gain. The confirmation consists of building the baseline and optimum combinations, with the same noise factors. These new boards are measured, S/N is calculated and compared to the predicted S/N and predicted Gain.

The purpose of the confirmation run is to ensure that there are no significant detrimental interactions, unknown noise factors or experimental error present. Confirmation provides confidence that the optimum combination is indeed more robust than the baseline combination.

21.3.2 Visual Scoring

The capability of the lead-free process was based on the grade of the solder joints that the process could consistently make. A good solder joint is one that connects the component to the circuit board and properly conducts current, shows sufficient mechanical strength, and is reliable over the life of the unit. The assumption was made that a solder joint that showed good wetting and soldering characteristics was a capable lead-free solder joint. The criteria for the grade of wetting were based on the industry standard IPC workmanship standards for SMT solder joints [7]. This system classifies joints based on visually observed wetting heights and widths of the solder on the component leads. In general, better wetting is shown by greater heights or lengths. Using this same concept of "higher the better," a visual scoring system was developed for grading the level of goodness of the lead-free solder joints that were made. This scoring system for the gull wing leaded parts is shown in Table 21.3.

The test vehicle selected was a 0.031 in. thick circuit board with FR4 laminate. The metalization of the board was nickel/gold (Ni/Au). The components studied included: an 8 pin SOIC (50 mil pitch) and an 0805

Table 21.3 Visual scoring criteria for gull wing joints

Score	Side overhang max	Toe overhang	Toe fillet height	Side Fillet	Heel fillet height, min	Shin height (top of foot)	Open space (between joints)
0 = Perfect	Lead does not overhang side of land	Toe does not overhang outer end of land	<= 100% of toe is wetted with proper angle	100% of side is wetted with proper angle	extends above lead thickness, does not fill upper lead bend, >200% of lead thickness	covers foot plus some of lead height	100% open
1 = Good	< 50 % of lead / land width	overhangs a bit, but easily meets min. electrical clearance	<75% of toe and good angle	<75% of side and good angle	>150% of lead thickness	some wetting on foot	<90% open
2 = Fair	50% of lesser of lead / land width	barely meets minimum electrical clearance	<50% of toe, and fair angle	<50% of side, and fair angle	>100% of lead thickness	no wetting on foot	<75% open
3 = NG marginal	slightly >50% of lead / land width	50-99% of minimum electical clearance	<50% of toe, bad angle	<50% of side, bad angle	< 100% of lead thickness	(not a rejectable criteria)	<50% open
4 = Bad	~75% of lead / land width	<25% of minimum electrical clearance	minimal height, some wetting	minimal height, some wetting	<50% of lead thickness		<25% open
5 = Horrible	100% off land	shorted to adjacent runner / component	zero height, no wetting	zero height, no wetting (pillow joint)	<25% of lead thickness or fills upper bend		0 % open (shorted)

resistor and an 0805 capacitor. These components were placed 5 times throughout the board, in each corner and once in the middle of the board.

For each combination in the L18 experiment, four arrays were built. For each array, 3 gull wing joints, 1 resistor and 1 capacitor joints were scored.

21.3.3 Pull Test

Another measure of the degree of wetting of solder joints was the force required to break the joint. Pull Testing was performed on the gull-wing solder joints. While the strength of a solder joint is greatly dependent on the alloy used, pull strength differences in the same component set could result if wetting mechanisms and/or resulting intermetallics are different due to the various flux types, printing, or reflow parameters [8]. The strength of a joint can also be considered an early life measure of the reliability of a joint. Since each DOE cell used the same alloy (within the tolerances of composition that the paste suppliers could supply), a larger force needed to break the joint was interpreted to mean a better solder joint.

For this experiment, the body package of the gull wing leaded component was cut away from the leads, resulting in a protruding lead from the circuit board. The top of the lead was clamped and pulled away from joint in the direction normal to the board. Once clamped, the joint was pulled at a rate of 0.8 mm/s until failure. The maximum force required to break the joint was recorded.

21.4 Results and Discussion

21.4.1 Visual Scoring Results

After calculating the S/N ratio for each run of the L18 experimental design, a response table could be assembled for the visual scoring system. In this response table, the average S/N ratio for each factor at each level can be determined. This response table for S/N visual scoring for all joints is shown in Table 21.4. This same data is presented graphically in Figure 21.2.

Figure 21.3 shows cross section photos of the best and worst runs of the L18 in terms of S/N ratio. These photos demonstrate how high S/N ratio corresponds to reduced variability. For run 10, the wetting heights of the joints at each noise level remain relatively consistent. Conversely, the wetting heights at the high noise level (which includes Pd/Ni/Au plated parts) for run 9 is significantly less than the wetting height at low noise level (which includes Sn plated parts).

Table 21.4 Response table for visual scoring S/N

Level	Paste brand	Print speed	Peak temp.	Time above liquidous	Conveyor speed	Nitrogen
1	4.1	5.3	6.0	5.6	4.9	7.7
2	5.8	5.3	4.7	5.9	6.2	6.5
3	6.7	6.4	6.4		5.9	2.8
4	7.0					
5	4.6					
delta	2.9	1.1	1.7	0.3	1.3	4.9
ranking	2	5	3	6	4	1

For each factor, the range of the S/N ratios among the different levels is shown in Table 21.4 as the delta. This delta value provides a measure of the relative level of influence that that particular factor has on the system output, in this case the visual score. The S/N is effected by the input factors in the following order:

N_2 Level > Paste Type > Peak Temp. > Conveyor Speed >> Print Speed > Time Above Liquidous

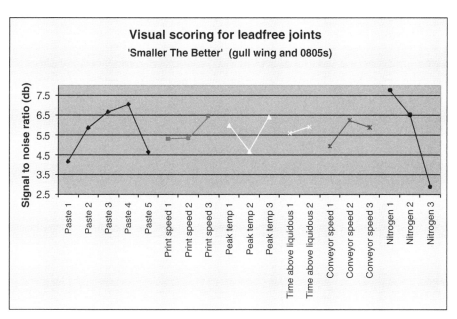

Figure 21.2 Response graph for visual scoring of wetting of lead-free joints

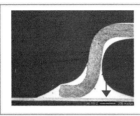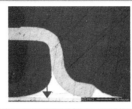

R6 - High S/N

Higher Signal/Noise ratio
resulted in both better wetting
and more consistent wetting
for high & low noise.

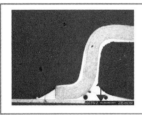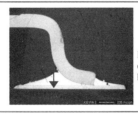

R16 - Low S/N

Lower S/N ratio resulted in
dissimilar wetting for high &
low noise.

Figure 21.3   Cross-section photos of best and worst S/N of L18 runs

Since the delta values for the first four factors are significantly greater than the values for the remaining two factors, only these first four are considered significant and will be considered in the model for best settings.

To choose the best settings for reducing variability of the lead-free soldering process with respect to noise, the setting for each factor with the largest S/N is chosen. For the solder paste type, the best S/N setting was not chosen due to the poor surface insulation resistance (SIR) performance of this paste. A paste with a better SIR value that still maintained good wetting was substituted.

The predicted S/N ratio that can be obtained from the optimal settings is calculated using the following formula:

$$S/N_{Optimal} = MaxS/N_{N2} + MaxS/N_{PasteType} + MaxS/N_{PeakTemp}$$
$$+ MaxS/N_{ConveyorSpeed} - 3 * \overline{T}$$

where $\overline{T}$ = average all S/N in Response Table.

The predicted S/N ratio for the baseline can be obtained from the following formula:

$$S/N_{Baseline} = BaseS/N_{N2} + BaseS/N_{PasteType} + BaseS/N_{PeakTemp}$$
$$+ BaseS/N_{ConveyorSpeed} - 3 * \overline{T}$$

where Base = Baseline Setting.

Table 21.5 Optimal and baseline settings for Pb-free process, and predicted and actual S/N for these settings

	PASTE BRAND	PRINT SPEED	PEAK TEMP	TAL	CONV. SPEED	N_2	Predicted S/N	Actual S/N
Selected "optimum" settings	5	3	3	2	2	1	8.0	7.0
Baseline settings	1	2	1	1	3	2	5.4	4.9
Gain							2.6	2.1

By comparing this optimal S/N ratio against the predicted S/N ratio for a baseline setting, it is possible to estimate the level of improvement to the lead-free system that the optimal settings can provide. For this system, the baseline was chosen to be typical settings for an SMT process, especially those that allow the line to run fastest and coolest. Table 21.5 shows the optimal settings and baseline settings for the visual scoring output. Table 21.5 also shows the predicted S/N for each of these settings, along with the process gain, which is the difference between optimal and baseline S/N values.

After the "optimal" process settings were selected, a confirmation run was performed with these settings. The baseline settings were also run at this time. The gull wing IC, resistors, and capacitors were again scored and S/N ratios again calculated. The actual S/N ratios for the optimal settings and baseline settings, and the actual gain, are also shown in Table 21.5.

From Table 21.5, it can be seen that the predicted S/N values and actual S/N values are very similar for both the optimal and baseline runs. Consequently, the predicted and actual process gains are also very similar. From this it can be concluded that the optimal settings are significantly better than the baseline.

The actual process gain can be related to percentage process variability reduction [10]. For a given gain, the percentage variability reduction of the system is related by the following equation:

$$\% VariationReduction = \left(1 - \frac{1}{2}^{\left(\frac{GAIN}{6} \right)} \right) * 100$$

For this lead-free system, it is predicted that the demonstrated gain of 2.1 dB corresponds to approximately a 22% reduction in the variation of the wetting of the solder joints, and therefore a corresponding increase in the overall quality of the lead-free process.

21.4.2 Pull Test Results

Pull testing was performed at "0 hours" on the individual gull wing IC leads for each of the runs of the L18. The same S/N analysis as was done with the visual scoring was performed for the pull test results. However, these results did not confirm. Not confirming can be due to: (1) strong interactions between control factors, (2) strong noise factors that were not studied and not controlled, (3) pure experimental error. Because of this lack of confirmation, this metric was not used to choose the optimum process settings.

While the pull test data could not be used to discern differences between the different process settings, it was able to show significant differences between the platting types and therefore lead-free metal systems used for ICs. The following three metal systems were compared:

- Tin/Lead Plating with Tin Lead Solder Alloy
- Tin Plating with SAC Solder Alloy
- Pd/Ni/Au Plating with SAC Solder Alloy.

Figure 21.4 shows the maximum force required to break the gull wing joints for each of these systems.

This data shows that the Pd/Ni/Au plating creates a weaker joint (with respect to normal stresses) than the Sn plating when using the SAC lead-free

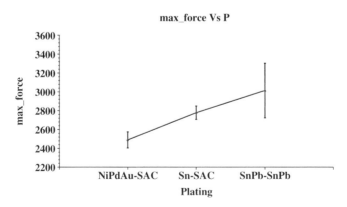

Figure 21.4 Maximum pull strength for leads with three different plating-alloy systems

alloy. The Sn plating with the SAC alloy creates a joint that is very similar in strength to a Sn/Pb plated part soldered with Sn/Pb alloy.

21.4.3 Next Steps

There are several steps planned for future work in this technology.

1. Conduct a robust optimization experiment like this one on a more complicated product. This will be done to extend the learning across a larger population of component types.
2. Evaluate reliability of the lead-free joints using the optimum and baseline combinations and compare the results to tin-lead results and to customer requirements.
3. Convert a current production part to the lead-free process and gain high volume production experience and field experience.

21.5 Conclusion

Nitrogen content in the reflow oven is the most significant factor when creating high quality lead-free joints. It is necessary to use highly inerted (low ppm oxygen) ovens to obtain lead-free solder joints that show the least variation when considering usual processing noise. The second most significant factor to create a robust lead-free manufacturing process is solder paste type. While there are a few different options, it is necessary to consider the tradeoff between those pastes that provide excellent wetting and those that have clean residues that are not a mitigating factor for electrochemical migration or dendritic growth. This is additionally true when considering lead-free joints since lead-free alloys tend not to wet as well as Sn/Pb alloys. This work demonstrates that there are solder pastes available that can provide both the needed wetting and the desirable level of residue cleanliness. Part plating metalization is a critical soldering process noise factor to consider when dealing with lead free. These results show Sn plated IC leads to create solder joints of equal strength to traditional Sn/Pb leads, while Ni/Pd/Au plating produced joints of lesser strength. Contributing to this reduction in joint strength is the reduced wetting that can be seen in the Pd/Ni/Au plated parts versus the Sn plated parts. Therefore, for a given product, it is necessary to consider the specific lead-free metal substitute used and whether that plating metal provides a system that can withstand the stresses for that product. A robust, lead-free SMT soldering process is feasible, and compensation for process noise, such as plating material, can be made by selecting the appropriate process parameters settings, especially for nitrogen level and paste type.

21.5.1 Acknowledgment

We would like to express appreciation to Jim Baar and Galen Reeder for sponsoring this Lead Free Project, Brad Walker and Rich Parker for their coaching and suggestions, Bob Clawson, Brian Chandler and Rick Graves for their help in assembly and data acquisition.

21.6 References

1. European Union (2002) Directive of the European Council and of the Parliament on the restriction of the use of certain hazardous substances in electrical and electronic equipment, PE-CONS 3662/02, November 8, 2002.
2. European Union (2002) Directive of the European Council and of the Parliament on waste electrical and electronic equipment, PE-CONS 3663/02, November 8, 2002.
3. Hunt, C., Lea, D., Adams, S., and Stratton, P. (2002) Evaluation of the comparative solderability of lead-free solders in nitrogen, part II, *Proceedings of the Technical Program of SMTA International, Chicago, IL, 2002.*
4. Klenke, B. (2002) *Lead-Free Selective Soldering: The Wave of the Future.* Plymouth, WI: ERSA, Inc.
5. Seelig, K. and Suraski, D. (2002) A study of lead-contamination in lead-free electronics assembly and its impact on reliability, *Proceedings of the Technical Program of SMTA International, Chicago, IL, 2002.*
6. Taguchi, G., Chowdhury, S., and Taguchi, S. (2000) *Robust Engineering.* New York: McGraw-Hill, p. 4.
7. IPC-A-610C, *Acceptability of Electronic Assemblies,* November 2001.
8. Henshall, G., Roubaud, P, Chew, G, *et al.* (2002) Impact of component terminal finish on the reliability of Pb-free solder joints, *Proceedings of the Technical Program of SMTA International, Chicago, IL, 2002.*
9. IPC/JEDEC J-STD-004A, *Requirements for Soldering Fluxes,* November 2000.
10. American Supplier Institute, *Robust Engineering Week 1 Workshop Manual,* 2002.

This case study is contributed by Fred Kuhlman and Mike Pepples of Delphi Delco Electronics Systems, USA and Craig Jensen of ASI Consulting Group, LLC, USA.

22

Catalyst Slurry Coating Process Optimization for Diesel Catalyzed Particulate Traps

Delphi Energy and Chassis Systems, USA

22.1 Executive Summary

A Diesel Particulate Filter (DPF) is an important component of a diesel exhaust system. The DPF minimizes particulates in the exhaust stream. Robust Engineering Parameter Design methods were applied to optimize the catalyst slurry coating process for the DPF. The initial production scale-up run was unsatisfactory with a process capability index (C_{pk}) less than 1.0. Three "Nominal-the-Best" Robust Engineering Parameter Design experiments were conducted on the process, one in Tulsa, Oklahoma and two in Florange, France. Very similar process improvements and learning's were obtained for each. Both Tulsa and Florange processes confirmed with 18 dB gains in Signal-to-Noise Ratios and on target performance. The C_{pk} index was greater than 2.

Robust Optimization: World's Best Practices for Developing Winning Vehicles,
First Edition. Subir Chowdhury and Shin Taguchi.
© 2016 Subir Chowdhury, Shin Taguchi, and ASI Consulting Group, LLC.
Published 2016 by John Wiley & Sons, Ltd.

22.2 Introduction

Diesel engines have widespread applications. They are used in ships, trucks, cars, busses, and heavy equipment, etc. Diesel engines have excellent reliability, durability, and fuel economy. Hydrocarbon and carbon monoxide emissions are low with low carbon dioxide. However, the public perceives diesel engines as intrinsically dirty. One has only to look at the black soot flowing from the exhaust to understand why.

Heavy trucks contribute approximately 18% of all vehicle contribution to air pollution and approximately 49% of nitric oxides (NOx) 62% of the particulate matter (PM). Diesel emissions of NOx and particulates have been steadily improving along with reductions in worldwide emissions standards.

Particulate traps (filters) will be required to meet future emissions standards. Particulate traps, shown in Figure 22.1, are porous honeycombed substrates with every other channel blocked. Exhaust gases flow into the open channels, through the porous wall and exit the adjacent channel. The particulates cannot pass through the wall and become trapped. Eventually, the backpressure will build up enough to degrade engine performance. The particulate trap must be regenerated at some point before this happens.

Figure 22.2 illustrates the build-up of backpressure until the regeneration threshold is reached, at which time an engine management event initiates the regeneration. Figure 22.2 also illustrates the difference between a noncatalyzed and a catalyzed particulate trap. The catalyzed trap takes longer to reach the regeneration threshold and regenerates faster.

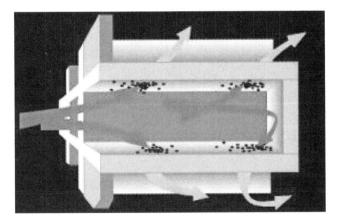

Figure 22.1   Ceramic particulate trap

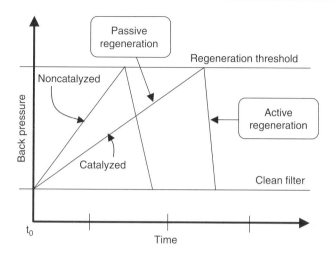

Figure 22.2   Particulate trap regeneration

The catalyst enhances passive and active regeneration by enabling the following reactions to occur at a lower temperature and at a faster rate:

$$C + O_2 \rightarrow CO_2$$
$$C + NO_2 \rightarrow CO + NO$$

The resulting constituents are handled downstream as required by worldwide emissions requirements.

22.3 Project Description

This is an advanced development project (ADP) that involves the development of a new generation catalyst slurry coating process for the Delphi Diesel Catalyzed Particulate Trap. It is imperative that the washcoat/catalyst be applied consistently from substrate to substrate and evenly within a substrate.

The objective is to optimize the catalyst slurry coating process to provide minimum variation in washcoat loading with on target performance, in the presence of raw material variation.

The process was developed in Tulsa, OK. Production will be in the manufacturing facility in Florange, France.

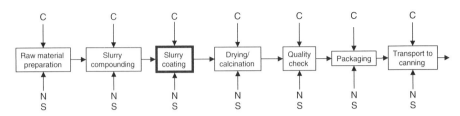

Figure 22.3   High level map of slurry coating process

22.4   Process Map

Figure 22.3 is a high level map of the slurry coating process. The process starts with raw material preparation, followed by slurry compounding. The slurry is transported to the coater where it is applied to the substrate. The coated substrate then goes to drying and calcination. After passing the quality checks, the diesel catalyzed particulate trap is packaged for transport the canning facility.

The slurry coating function was selected for Parameter Design Optimization based on 1 – it is a new process, 2 – it is sensitive to variation in raw materials, and 3 – process FMEA results.

22.4.1   Initial Performance

The initial ADP production scale-up run in Florange, France highlighted significant opportunity for improvement in process performance, see Figure 22.4. The requirement for slurry loading is 121.6 ± 12 grams. The process was running at 146.6 grams with a standard deviation of 19 grams. The C_{pk} was less than 1.

A clear difference in process performance was observed when the part measurements were plotted in production order. This difference in performance was due to lot-to-lot variation in incoming raw material. The variation in raw material was included in the first Parameter Design experiment as Noise Factor Q.

22.5   First Parameter Design Experiment

The first Parameter Design experiment was conducted in Florange, France with support from a process engineer from Tulsa, OK.

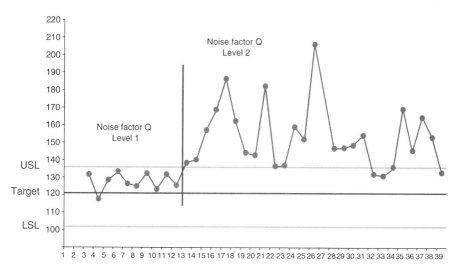

Figure 22.4 Initial production scale-up run

22.5.1 Function Analysis

Process maps and function analysis techniques are powerful tools to gain understanding of the process operation and to help determine which factors should be included in a Parameter Design Experiment. Figure 22.5 is the slurry coating process model. The slurry from "slurry compounding" is introduced into a slurry mixing tank where it undergoes further processing. The coater is charged with slurry and a substrate is loaded into the coater. The substrate is then coated. "Coat Substrate" was determined to be a critical function for the reasons stated previously.

The "Coat Substrate" function was further analyzed with the function analysis depicted in Figure 22.6. When function analysis diagrams are read left to right, they represent "how" the function the immediate left is accomplished. The analysis diagram must make sense when read right to left also. When read right to left, the diagram indicates "why" the function to the immediate right is performed.

For example reading from left to right, how we "Coat Substrate" is to "Apply Slurry into Substrate" and reading from right to left, why we "Apply Slurry into Substrate" is to "Coat Substrate."

The parameters are the factors that are candidates to be included in a Parameter Design experiment. The boxes with (?) represent the parameter values that are determined by the Parameter Design Experiment.

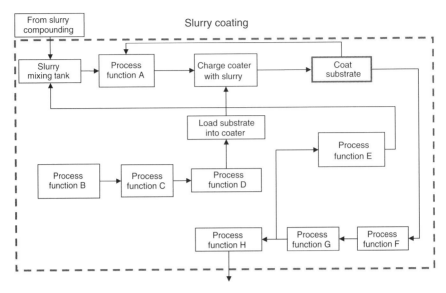

Figure 22.5   Slurry coating process map

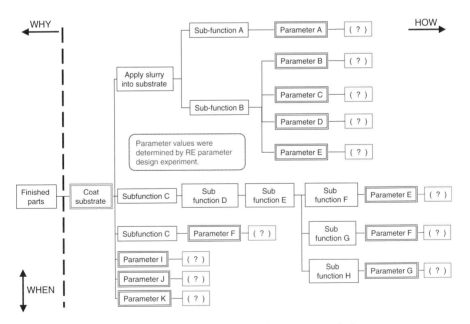

Figure 22.6   Slurry coating function analysis

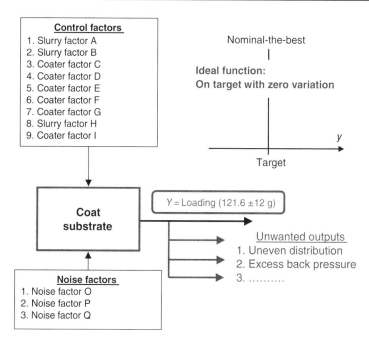

Figure 22.7   Parameter diagram

22.5.2 Ideal Function

The Ideal Function in shown in Figure 22.7. The slurry coating process was being optimized to coat the substrate with 121.6 grams of washcoat. The Nominal-the-Best Signal-to-Noise Ratio was used. The S/N was calculated as

$$S/N = 10*\log\left(\frac{\overline{Y}}{s}\right)^{2}$$

22.5.3 Measurement System Evaluation

Ensuring an adequate measurement system is critical for any experiment. A Parameter Design experiment is no exception. Table 22.1 is the measurement system evaluation (MSE) for this project. It should be noted that the objective is not to certify a gage for production. The objective is to evaluate the measurement system for the particular development project.

Table 22.1 Measurement system evaluation

Measurement Variation

Sample	$y_{i(m1)}$	$y_{i(m2)}$	$\bar{y}_{sample}$	$e_{i(mj)} = y_{i(mj)} - \bar{y}_{sample}$	
1	3266.3	3266.1	3266.200	0.1	−0.1
2	3267.1	3266.8	3266.950	0.15	−0.15
3	3267.3	3267.1	3267.200	0.1	−0.1
4	3270.3	3270.5	3270.400	−0.1	0.1
5	3270.9	3270.8	3270.850	0.05	−0.05
6	3275.4	3275.2	3275.300	0.1	−0.1
7	3282.5	3282.5	3282.500	0	0
8	3283.5	3283.5	3283.500	0	0
9	3290.0	3289.9	3289.950	0.05	−0.05
10	3292.0	3291.8	3291.900	0.1	−0.1
$\bar{y}_{mj} = 3276.530$		3276.420	$\sum e^2_{i(mj)} = 0.155$		
$s_{mj} = 9.750$		9.772	$s_m = \sqrt{\dfrac{\sum e^2_{i(mj)}}{df}}, df = 10\ 0.124$		

Ten samples were each measured twice. The measurement system standard deviation, s_m was calculated using the equations in Table 22.1.

The minimum detectable effect, at an approximate 95% confidence level, for the experiment is calculated from the following equation:

$$d = \left(8s_m \big/ \sqrt{n}\right)$$

where:

d = the minimum detectable effect for the experiment;
s_m = the measurement system standard deviation;
n = the number of treatment combinations in the experiment.

The planned Parameter Design experiment was an L_{12} with N1 and N2 noise conditions in the outer array. The total number of treatment combinations is $2 \times 12 = 24$. The minimum detectable effect is $d = (8 * 0.124 / \sqrt{24}) = 0.20g$. The measurement system is adequate for the experiment since the effects of interest are much larger.

22.5.4 Parameter Diagram

Figure 22.7 shows the Parameter Diagram and the Ideal Function. The function to be optimized is "Coat Substrate." There are three slurry control factors and six coater control factors. Of the three noise factors, noise factor Q was added by the process engineer at the time of the experiment. The Ideal Function is to be on target with zero variation. The target is 121.6 grams of loading. There is a tolerance of ±12 grams. Some of the unwanted outputs of this process are loading variation from part to part, uneven distribution within parts, and excess back pressure.

22.5.5 Factors and Levels

Table 22.2 and Table 22.3 document the control and noise factors and their levels. The nosie strategy was changed by the process engineer based on the findings from the initial production scale-up run. Lot-to-lot variation was used as noise factor Q. Ordinarily, lot-to-lot variation is not a good noise factor. However, it was used as a surrogate for the true, but unknown noise factor. Follow-up engineering investigation determined the true noise factor, which was used in subsequent Parameter Design experiements.

Table 22.2   Control factors

Control factors	Level 1	Level 2
A. Slurry factor A	0.5	1.5
B. Slurry factor B	10.00	20.00
C. Coater factor C	50	100
D. Coater factor D	60	120
E. Coater factor E	0	4
F. Coater factor F	60	120
G. Coater factor G	40	70
H. Slurry factor H	0	0.2
I. Coater factor I	1	2

Table 22.3   Noise factors

Noise factors	Level 1	Level 2
1. Noise factor O	Low	High
2. Noise factor P	−2	+2
3. Noise factor Q	832	963

Table 22.4   Noise factors levels for N_1

Noise conditions when the response tends to be lower	
Noise factor	Level
1. Noise factor O	Low
2. Noise factor P	−2.0
3. Noise factor Q	963

Table 22.5   Noise factors levels for N_2

Noise conditions when the response tends to be higher	
Noise factor	Level
1. Noise factor O	High
2. Noise factor P	+2.0
3. Noise factor Q	832

22.5.6   Compound Noise Strategy

The purpose of a compound noise strategy, Table 22.4, Table 22.5 and Figure 22.8, is to force performance away from the nominal value. N1 noise conditions force performance below nominal and N2 noise conditions force performance above nominal. This will facilitate learning how to reduce variation in the presence of noise.

22.5.7   Parameter Design Experiment Layout (1)

Table 22.6 is the Parameter Design experiment layout for the initial experiment performed in Florange, France. An L_{12} orthogonal array was selected for

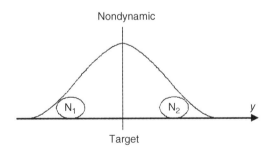

Figure 22.8   Compound noise strategy

Table 22.6 Parameter design layout (1)

tc	Factor A A	Factor B B	Factor C C	Factor D D	Factor D E	Factor F F	Factor G G	Factor H H	Factor I I	Loading N1	Loading N2	Ybar	Stdev	S/N
1	0.50	10.00	50.00	60.00	0.00	60.00	40.00	0.00	1	38.9	40.9	39.90	1.44	28.84
2	0.50	10.00	50.00	60.00	0.00	120.00	70.00	0.20	2	41.1	43.0	42.07	1.38	29.69
3	0.50	10.00	100.00	120.00	4.00	60.00	40.00	0.20	2	120.7	122.6	121.68	1.35	39.09
4	0.50	20.00	50.00	120.00	4.00	60.00	70.00	0.00	1	92.2	103.6	97.88	8.03	21.72
5	0.50	20.00	100.00	60.00	4.00	120.00	70.00	0.00	2	324.6	335.6	330.12	7.75	32.59
6	0.50	20.00	100.00	120.00	0.00	120.00	40.00	0.20	1	65.5	104.5	84.97	27.56	9.78
7	1.50	10.00	100.00	120.00	0.00	60.00	70.00	0.00	2	75.7	93.9	84.82	12.90	16.36
8	1.50	10.00	100.00	60.00	4.00	120.00	40.00	0.00	1	89.6	124.4	107.01	24.57	12.78
9	1.50	10.00	50.00	120.00	4.00	120.00	70.00	0.20	1	44.6	52.2	48.36	5.39	19.06
10	1.50	20.00	100.00	60.00	0.00	120.00	70.00	0.20	1	172.4	174.9	173.68	1.78	39.78
11	1.50	20.00	50.00	120.00	0.00	120.00	40.00	0.00	2	115.0	126.2	120.63	7.91	23.67
12	1.50	20.00	50.00	60.00	4.00	60.00	40.00	0.20	2	111.1	125.7	118.41	10.28	21.23
									Process S/N from initial production scale-up run data					
Int	1.25	16.27	85.00	90.00	0.00	90.00	70.00	0.00	2	NA	NA	146.61	19.2	17.66

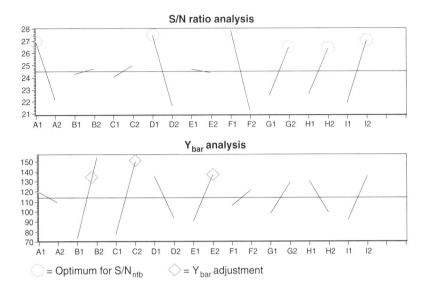

Figure 22.9 Means plots

the nine control factors. The information from the initial production scale-up run is presented at the bottom of the array.

The data for noise condition N2 of treatment combination 6 was missing. The S/N was set to 9.78 dB (3 dB below the lowest S/N) and the value of N2 was calculated using the Excel Goal Seek function.

It was determined that control factor H at level 2 was the cause of the missing data. This information was used in subsequent Parameter Design Experiments.

22.5.8 Means Plots

The means plots for S/N Ratios and $\bar{y}$ in Figure 22.9 are a dream come true for Robust Engineering practitioners. There are six control factors that have a strong affect on S/N. The three factors that have minimal affect on S/N have a strong affect on $\bar{y}$ and can be used to adjust the average performance to target.

22.5.9 Means Tables

The means tables, Table 22.7a and Table Table 22.7b, are presented to support the means plots and the following two-step optimization and prediction.

Table 22.7 Means tables: (a) S/N ratio; (b) Y_{bar}

Means Table - Signal to Noise Radio $S/N_{bar} = 24.51$

	A	B	C	D	E	F	G	H	I
Level1	26.91	24.27	24.02	27.46	24.64	27.82	22.50	22.63	21.93
Level2	22.11	24.75	25.00	21.56	24.38	21.20	26.52	26.39	27.09
Delta	4.80	0.48	0.97	5.90	0.25	6.63	4.01	3.76	5.17
Rank	4	8	7	2	9	1	5	6	3

(a)

Means Table - Y_{bar} $T_{bar} = 114.13$

	A	B	C	D	E	F	G	H	I
Level1	119.44	73.97	77.87	135.20	91.01	106.06	98.77	130.06	91.97
Level2	108.82	154.28	150.38	93.06	137.24	122.19	129.49	98.19	136.29
Delta	10.62	80.31	72.50	42.14	46.23	16.13	30.72	31.87	44.32
Rank	9	1	2	5	3	8	7	6	4

(b)

22.5.10 Two-Step Optimization and Prediction

Step 1 – Maximize S/N. Control factor levels A1, D1, F1, G2, H2, and I2 were selected to maximize S/N. Control factors B, C, and E can be used adjust the process mean to target, based on cost and manufacturing considerations.

The predicted performance is:

$$S/N = \overline{A}_1 + \overline{D}_1 + \overline{F}_1 + \overline{G}_2 + \overline{H}_2 + \overline{I}_2 - (5 * S/N_{bar})$$
$$S/N = 39.64\, dB$$
$$\overline{y} = 121.6\, g\, (\text{adjusted to target})$$
$$stdev = 1.27$$

The predicted process standard deviation was calculated by manipulating the S/N equation as follows:

$$S/N = 10 \log \left(\frac{\overline{y}}{s}\right)^2$$

$$s = \left(\frac{\overline{y}}{10^{\left(\frac{S/N}{20}\right)}}\right)$$

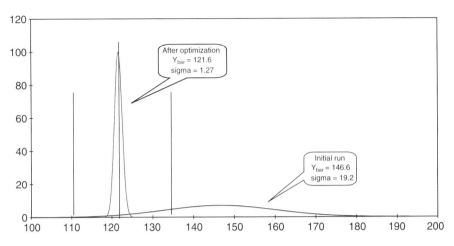

Figure 22.10 Performance before and after optimization

22.5.11 Predicted Performance Improvement Before and After

Figure 22.10 depicts the predicted process performance before and after optimization. σ is predicted to be reduced from 19.2 g to 1.27 g. This is a reduction in σ of approximately 93%. $\bar{y}$ is adjusted from the initial value of 146.6 g to be on target at 121.6 g. This is a remarkable improvement. However, it must be remembered that control factor H at level 2 caused a problem and was not a viable manufacturing option.

22.6 Follow-up Parameter Design Experiment

Given that control factor H at level 2 was not a viable manufacturing option; a follow-up Parameter Design experiment was performed on a similar substrate coater in Tulsa, OK, to further optimize the coating process. The P – Diagram and Ideal Function are shown in Figure 22.11.

This was a small experiment with only two coater control factors. The slurry control factor H was set at level 1 and all other slurry control factors were held constant. Noise factor Q is now the ture noise factor that was causing the lot-to-lot variation in the previous experiment. The target loading remains the same.

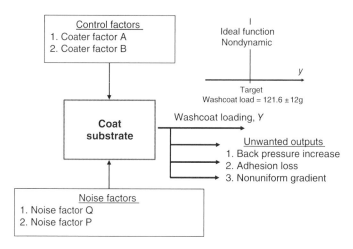

Figure 22.11   P-diagram, Tulsa

22.6.1   Parameter Design Experiment Layout (2)

The Parameter Design experiment layout is given in Table 22.8. This is a small experiment with an L_4 inner array and an L_4 in the outer array. The noise factors were not compounded for this experiment. Again, the true noise factor causing the lot-to-lot variation observed in the first experiment is used for noise factor Q.

22.6.2   Means Plots for Signal-to-Noise Ratios

The means plots for signal-to-Noise Ratios (Figure 22.12) indicate that level 1 is preferred for maximum S/N. The predicted S/N is 37.8 dB with on target performance of 121.6 grams of slurry loading. σ is reduced by approximately 85%.

22.6.3   Confirmation Results in Tulsa

The confirmation results for the follow-up Parameter Design experiment in Tulsa, OK are shown in Figure 22.13. This figure indicates on target perform-ance with $\sigma = 1.56$ grams.

22.6.4   Noise Factor Q Affect on Slurry Coating

The affect of noise factor Q on process control is clearly evident on Figure 22.14. Noise factor Q had a significant effect on process performance prior to optimization. It affected both the process mean and variability. The affect on process performance after optimization has been minimized. Note that the scale is different.

Table 22.8 Parameter design layout (2)

tc	factor A	factor B	Noise factor Q - Low		Noise factor Q - High		Ybar	Stdev	S/N
			Noise factor P - Low	Noise factor P - High	Noise factor P - Low	Noise factor P - High			
1	1	1	136.4	141.0	138.3	141.2	139.23	2.30	35.64
2	1	2	200.2	234.8	204.3	241.5	220.20	20.97	20.42
3	2	1	124.5	135.7	129.6	137.3	131.78	5.88	27.01
4	2	2	181.3	197.8	194.0	198.2	192.83	7.91	27.74

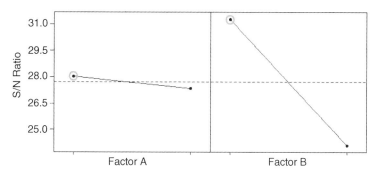

Figure 22.12   Means plots for S/N

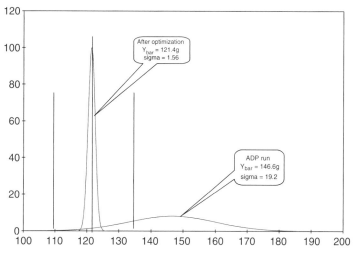

Figure 22.13   Confirmation results in Tulsa

22.7 Transfer to Florange

Figure 22.15 is the results of the process capability study conducted in Florange, France. The study was conducted with factor settings from the follow-up Parameter Design experiment in Tulsa, OK. The process capability is clearly unsatisfactory, with a C_{pk} less than 1.0. However, the standard deviation of 8 grams is still much less than the standard deviation of 19 grams from the initial production scale-up run.

The unsatisfactory performance was due to new target loading requirements caused by changes to the substrate volume and the installation of a second-generation coater.

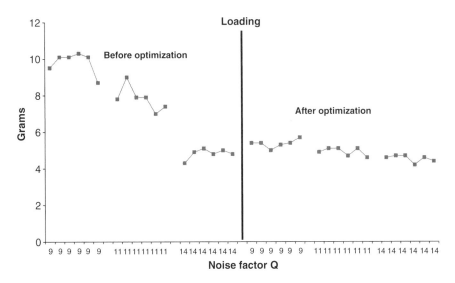

Figure 22.14   Noise factor Q affect on process control

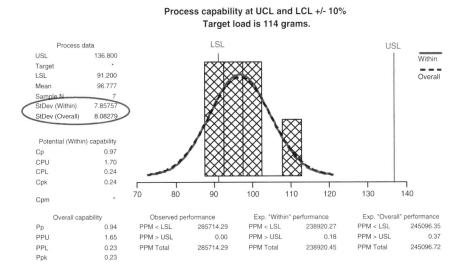

Figure 22.15   Process capability study: Tulsa settings

The project team elected to run another Parameter Design experiment, since the control factor settings were not directly applicable to the second-generation coater.

22.7.1 Ideal Function and Parameter Diagram

The basic Parameter Diagram and Ideal Function for both manufacturing locations (Tulsa and Florange) are shown in Figure 22.16. The slurry loading target changed from 121.6 ± 12 to 114 ± 11 grams to accommodate the changes in the substrate volume and the second-generation coater. The slurry control factors remained constant. The coater control factors and levels were selected based on knowledge gained from the first Parameter Design experiment.

The noise factors are the same as the first and second experiment. As in the second Parameter Design experiment, the true noise factor causing lot-to-lot variation in raw material was used as Noise Factor Q.

22.7.2 Parameter Design Experiment Layout (3)

Table 22.9 is the Parameter Design experiment layout. An L_{18} array was selected with seven coater control factors. The "Dummy Treatment" method was applied to factor E.

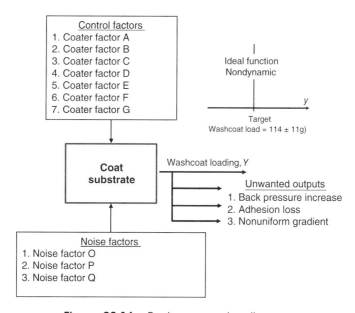

Figure 22.16 Basic parameter diagram

Table 22.9 Parameter design layout (3)

tc	Factor A	Factor B	Factor C	Factor D	Factor E	Factor F	Factor G	Loading				
	A	B	C	D	E	F	G	N1	N2	Ybar	stdev	S/N
1	outlet	100	100	85	30	0.6	10	101.1	108.3	5.09117	104.7	26.26
2	outlet	250	400	120	30	0.6	30	132.2	140.1	5.58614	136.2	27.74
3	outlet	400	250	120	77	0.6	60	90.7	125.8	24.8194	108.3	12.79
4	inlet	100	400	105	77	0.6	10	87.2	108.6	15.1321	97.9	16.22
5	inlet	250	100	105	77	0.6	60	83.6	91.5	5.58614	87.6	23.90
6	inlet	400	250	85	77	0.6	30	67.1	75.9	6.22254	71.5	21.21
7	outlet	100	250	105	77	1.2	30	101.2	105.1	2.75772	103.2	31.46
8	outlet	250	100	85	77	1.2	60	67.0	83.3	#DIV/0!	83.3	#DIV/0!
9	outlet	400	400	85	77	1.2	10	85.3	87.5	1.55563	86.4	34.89
10	inlet	100	100	120	77	1.2	30	89.5	98.1	6.08112	93.8	23.76
11	inlet	250	250	120	30	1.2	10	78.8	93.9	#DIV/0!	93.9	#DIV/0!
12	inlet	400	400	105	30	1.2	60	72.3	85.2	#DIV/0!	85.2	#DIV/0!
13	outlet	100	400	120	77	1.0	60	138.2	152.5	10.1116	145.4	23.15
14	outlet	250	250	105	77	1.0	10	98.8	109.8	7.77817	104.3	22.55
15	outlet	400	100	105	30	1.0	30	111.7	139.0	19.304	125.4	16.25
16	inlet	100	250	85	30	1.0	60	77.0	92.7	11.1016	84.9	17.67
17	inlet	250	400	85	77	1.0	30	63.7	70.1	#DIV/0!	70.1	#DIV/0!
18	inlet	400	100	120	77	1.0	10	90.7	97.7	4.94975	94.2	25.59

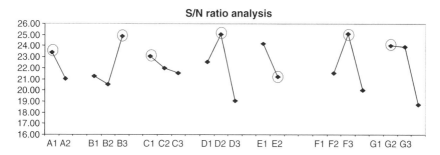

Figure 22.17   Means plots for S/N ratios

22.7.3 Means Plots for Signal-to-Noise Ratios

The means plots for the Signal-to-Noise Ratios, Figure 22.17, suggest that setting the coater control factors to A1, B3, C1, D2, E1, F2, and G1 will optimize the S/N ratio. It was decided to sacrifice some robustness by setting factor E to level 2 and for manufacturing considerations.

22.7.4 Prediction and Confirmation

Table 22.10 shows the prediction and confirmation results. The confirmation results compare favorably with the predicted performance. The S/N improvement is 17.5 dB. The reduction in σ from 19.2 grams in the initial production scale-up run to 1.8 grams in the confirmation run is approximately 90%.

22.7.5 Process Capability

The process capability study based on the confirmation run is shown in Figure 22.18. The C_{pk} is greater than 2. It will be a six sigma process. This

Table 22.10   Prediction and confirmation

Configuration	Y_{bar}	S/N	s
Initial	100.12	18.52	11.87
Predicted	114.10	33.47	2.42
Confirmed	114.02	35.99	1.81
Gain		17.47	

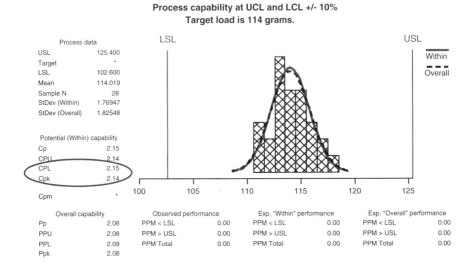

Figure 22.18   Process capability study – Florange settings

performance is in the presence of a noise strategy that represents variation in raw materials and expected variation of other process inputs. It should be a much better indicator of future production performance than a traditional short run capability study.

22.8 Conclusion

This was an advanced development project to develop a next generation catalyst slurry coating process for the Delphi Diesel Catalyzed Particulate Trap. Three Parameter Design experiments were conducted to this end. Each experiment was based on learning's from the preceding experiment. An approximate 90% reduction in σ was realized. The process is on target with a C_{pk} greater than 2.0. The improvements were accomplished despite changes in the target loading.

22.8.1   The Team

Contributions from a broad cross-functional team are required for the completion of any project. Robust Engineering projects are no exception. The authors are indebted to the following team members, whose assistance made this project possible:

John Fuerst	Business Line Executive
Hank Sullivan	Chief Engineer
Danan Dou	Engineering Manager
John Yan	Product Scientist
Laurent Faillon	Manufacturing Engineer
Serge Aldon	Test Engineer
Pat Canning	Robust Engineering Practitioner Candidate

This case study is contributed by Michael D. Holbrook and Justin P. Volpe of Delphi Energy and Chassis Systems, USA.

Index

Note: Page numbers in italic refer to illustrations

Robust Optimization: World's Best Practices for Developing Winning Vehicles,
First Edition. Subir Chowdhury and Shin Taguchi.
© 2016 Subir Chowdhury, Shin Taguchi, and ASI Consulting Group, LLC.
Published 2016 by John Wiley & Sons, Ltd.